菌草技术

JUNCAO TECHNOLOGY

林占熺　林冬梅 ◎ 著
Lin Zhanxi　Lin Dongmei

图书在版编目（CIP）数据

菌草技术 / 林占熺，林冬梅著. —福州：福建科学技术出版社，2022.8

ISBN 978-7-5335-6520-6

Ⅰ.①菌… Ⅱ.①林…②林… Ⅲ.①食用菌-蔬菜园艺②药用菌类-栽培技术 Ⅳ.①S646②S567.3

中国版本图书馆CIP数据核字（2021）第153888号

书　　名	菌草技术
著　　者	林占熺　林冬梅
出版发行	福建科学技术出版社
社　　址	福州市东水路76号（邮编350001）
网　　址	www.fjstp.com
经　　销	福建新华发行（集团）有限责任公司
印　　刷	福建新华联合印务集团有限公司
开　　本	889毫米×1194毫米　1/16
印　　张	13
字　　数	309千字
版　　次	2022年8月第1版
印　　次	2022年8月第1次印刷
书　　号	ISBN 978-7-5335-6520-6
定　　价	169.00元

书中如有印装质量问题，可直接向本社调换

习近平向菌草援外20周年暨助力可持续发展国际合作论坛致贺信

9月2日,国家主席习近平向菌草援外20周年暨助力可持续发展国际合作论坛致贺信。

习近平指出,菌草技术是"以草代木"发展起来的中国特有技术,实现了光、热、水三大农业资源综合高效利用,植物、动物、菌物三物循环生产,经济、社会、环境三大效益结合,有利于生态、粮食、能源安全。我长期关心菌草技术国际合作,2001年中国首个援外菌草技术示范基地在巴布亚新几内亚建成落地,至今这一技术已推广至全球一百多个国家,合作紧扣消除贫困、促进就业、可再生资源利用和应对气候变化等发展目标,为促进当地发展和人民福祉发挥了重要作用,受到发展中国家普遍欢迎。

习近平强调,中国愿同有关各方一道,继续为落实联合国2030年可持续发展议程贡献中国智慧、中国方案,使菌草技术成为造福广大发展中国家人民的"幸福草"!

菌草援外20周年暨助力可持续发展国际合作论坛当日在北京以线上线下相结合方式举行,由国家国际发展合作署和福建省人民政府共同举办。

引自2021年9月2日新华社电讯

Xi sends congratulatory letter to Forum on 20th Anniversary of Juncao Assistance, Sustainable Development Cooperation

Chinese President Xi Jinping on Thursday sent a congratulatory letter to the Forum on the 20th Anniversary of Juncao Assistance and Sustainable Development Cooperation, which was held both online and offline in Beijing.

In his letter, Xi noted that Juncao technology is a unique technology developed in China by "substituting grass for wood". It has realized the comprehensive and efficient utilization of the three major agricultural resources of light, heat and water, and attained cycle production of plants, animals and fungus. The technology generates economic, social and environmental benefits, and helps improve ecology, food and energy security, he said. "I have long cared about the international cooperation of Juncao technology," said Xi, adding that in 2001, the first China-aided Juncao technology demonstration base overseas was established in Papua New Guinea. So far, this technology has been expanded to more than 100 countries. He said that the international cooperation of Juncao technology closely followed development goals such as poverty eradication, employment promotion, renewable resource utilization, and dealing with climate change. This cooperation initiative has played an important role in promoting local development and people's well-being and is generally welcomed by developing countries, Xi added.

Xi emphasized that China is willing to work with relevant parties to continue to contribute China's wisdom and China's solutions to the implementation of the United Nations 2030 Agenda for Sustainable Development and make Juncao technology a "grass of happiness" that benefits people in developing countries.

前言

菌草技术是为扶贫和保护生态环境从"以草代木"栽培食药用菌发展起来的中国特有技术。

20世纪70年代前，中国的香菇、木耳、灵芝等食药用菌生产以阔叶树木材为原料，菌业生产与林业生态平衡产生"菌林矛盾"。为解决这一难题，著者（之一）从1983年开始用野生草本植物和农作物秸秆"以草代木"栽培食药用菌的研究，1986年用芒萁、五节芒、菅、类芦、斑茅、芦苇等野生草本植物栽培食药用菌获得成功，发明了菌草技术，为菌业生产开辟了可持续发展的新途径。

经过近40年的持续创新，菌草技术的研究与应用从"以草代木"栽培食药用菌拓展到"以草代粮"发展畜业，"以草代煤"发电，"以草代木"生产纤维板、纸浆生态治理生物肥料、等方面，开辟了"菌"与"草"交叉的科学研究与产业发展新领域。

科学研究进展

第一，菌草品种选育。选育出一批高固碳、高固氮、高光效、抗逆性强、生产力高、营养丰富的系列草种，成为新型的农业资源和生物材料。目前广泛栽培应用的巨菌草（*Cenchrus fungigraminus*）是蒺藜草属（*Cenchrus*）的新种，为多年生或一年生C4植物，具有高生物量、高蛋白、抗逆性强、适应性广等优良特性。

第二，"以草代木"栽培食药用菌。已筛选培育出适用菌草栽培的15个科56个品种的食药用菌，不仅可完全替代木屑栽培食用菌，而且具有产量高、质量好、可循环综合利用等特点。用菌草栽培的药用菌其有效药用成分高于用林木栽培的，可以从根本上解决"菌林矛盾"，实现可持续发展。

第三，"以草代粮"发展畜业。1990年，开始利用菌草栽培香菇的菌糟替代部分精饲料养猪的研究。此后20多年，先后在菌草饲料、菌草菌物饲料、菌草菌糟饲料、菌草食药用菌的菌糟提取物作饲料添加剂等方面开展研究，取得系列成果，形成"植物—菌物—动物"三物良性循环生产的新的生产方式和产业发展模式。菌草栽培灵芝的菌糟作为奶牛的饲料还可提高奶牛免疫力。

第四，发展生物质能源。2008年6月，和浙江省兰溪市热电厂合作，利用巨菌草发电，实测结果每公顷产的巨菌草燃烧发电量相当于52.5～60吨原煤的发电量。

第五，"以草代木"作为生物材料。该研究始于2009年，包括用巨菌草、绿洲1号等菌草品种来生产人造板、制浆造纸、制备活性炭等。巨菌草、绿洲1号等是工业开发的新的优质生物材料。由于产量高、适应性强，菌草用于生物材料的发展潜力巨大、前景广阔。

第六，菌草用于生态治理。自1993年以来，在福建、重庆、云南、新疆、西藏、贵州等省（自治区、直辖市），以及沿黄河九省，开展菌草治理水土流失、治理崩岗、治理荒漠、防风固沙、治理砒砂岩、治理洪积扇、滨海防风固沙、改良盐碱地的研究与示范，攻克了沿黄河生态脆弱区菌草生态屏障建设与产业发展的关键技术问题。2011～2021年，在非洲尼罗河源头的卢旺达鲁伯纳，中国援卢旺达农业技术示范中心开展种植巨菌草治理水土流失试验。2012年10月30日，在2.5小时内降雨量达51.4毫米的情况下，实验区的降雨全部被巨菌草植被截留。当地种植巨菌草与传统作物玉米比较，土壤流失减少97%以上。巨菌草与玉米、甘薯、黄豆等作物间作，简单易行，见效快、效果好、经济效益高。在国内外，菌草用于生态治理取得系列成果，形成菌草生态治理技术体系，开创了生态治理新途径。

菌草技术推广应用

1987年5月，菌草技术开始在中国福建闽西北贫困地区应用。取得成效后，菌草技术被福建省列为科技兴农、为民办实事、对口帮扶宁夏、智力援疆、智力援藏项目，被国家科委列为国家级星火计划重中之重项目，被中国扶贫基金会列为科技扶贫开发首选项目。至今，菌草技术已在中国31个省（自治区、直辖市）500多个县推广应用，为脱贫攻坚作出积极贡献。1994年起，菌草技术被中国政府列为援助发展中国家培训项目，被联合国开发计划署列为中国与其他发展中国家优先合作项目。2017年5月，菌草技术被联合国列为"中国—联合国和平与发展基金"重点推进项目。联合国经社部认为，菌草技术可有效促进落实"联合国2030年可持续发展"17个目标中的13个目标，包括消除贫困、粮食安全、健康、教育、性别平等、清洁能源、可持续经济增长消费和生产、创新、减少不平等、生态环境保护、应对气候变化、防治荒漠化和保护生物多样性和构建伙伴关系等。通过援外和国际合作，菌草技术已传播到106个国家，在巴布亚新几内亚、斐济、莱索托、卢旺达、中非、南非、马达加斯加、尼日利亚、纳米比亚、马来西亚、缅甸、老挝、泰国、朝鲜、巴西等16个国家建立菌草技术示范培训中心（基地），为国际减贫事业和落实联合国2030年可持续发展议程提供"中国方案"。

2021年9月2日，由国家国际发展合作署和福建省人民政府在北京联合举办的菌草援外20周年暨助力可持续发展国际合作论坛，国务委员兼外交部部长王毅宣读国家主席习近平贺信并作主旨讲话。联合国秘书长古特雷斯向论坛发来贺函，巴布亚新几内亚总理马拉佩、中非总统图瓦德拉、巴基斯坦总理伊姆兰、老挝总理潘坎、太平洋岛国论坛秘书长普纳等分别视频致贺。大会有力地推动世界菌草事业的发展，未来世界各地菌草技术更加广泛的应用将在一定范围内改变农业的生产方式和产业发展模式，大幅度提高成千上万贫困农民的生活水平，为构建人类命运共同体起积极作用。

为适应世界各国菌草事业发展的需要，我们把菌草技术研究及其应用的主要成果整理成此书。为让读者能直观了解菌草技术及应用概况，本书以图片为主，全书共收录图片355幅，以中英文对照的方式呈现。

在本书撰写过程中，承蒙有关院士、专家的悉心指导和各位同仁的支持帮助，在此谨表谢意，同时向所有关心、帮助菌草事业的领导和从事菌草事业的科技工作者和生产者表示衷心的感谢！

林占熺，林冬梅
2022年3月

Preface

Juncao technology is a unique technology originated in China to cultivate edible and medicinal fungi for poverty alleviation and environment protection.

Before the 1970s, the production of edible and medicinal fungi such as *Lentinusedodes*, *Auricularia auricular*, and *Ganoderma lucidum* in China took the wood of broad-leaf trees as the main substrate material which resulted in a prominent contradiction between fungi cultivation and forest. To crack this nut, the author started from 1983 to conduct research on substituting wood with wild herbs and crop straws to grow edible and medicinal fungi. In 1986, *Dicranopteris dichotoma*, *Miscanthus floidulus*, *Themeda gigantean* var. *villosa*, *Neyraudia reynaudiana*, *Saccharum arundinaceum* were used to cultivate edible and medicinal fungi, and he succeeded in his experiment and invented Juncao technology, which opened up a new pathway for the sustainable development of fungi production.

With nearly 40 years of unremitting research, besides substituting for wood to grow edible and medicinal fungi, Juncao grass now can be widely used in environment protection and fertilizer production. The application of Juncao grass has expanded to replace grain to develop animal husbandry, replace coal to generate electricity, wood to produce fiberboard and make paper. Juncao technology has opened up a new field of scientific research and industrial where mushroom production is combined with grass planting.

Progress of scientific research

First, the breeding of Juncao varieties. A series of grass species with high carbon and nitrogen fixation capacity, high photosynthetic rate, strong stress resistance, high productivity and rich nutrition have been selected and bred. It has become a new type of agricultural resource and biological material. *Cenchrus fungigraminus*, which is widely cultivated and applied at present, is a new species of the *Cenchrus* family. It is a perennial or annual C4 plant with excellent characteristics such as high biomass, high protein, resistance to adversity and wide adaptability.

Second, replacing wood with Juncao grass to cultivate edible and medicinal mushrooms. A total of 56 varieties of 15 families have been screened and cultivated. The mushrooms cultivated with Juncao grass have high yield, good quality, and enable recyclable and comprehensive utilization of resources. The mushroom cultivated with Juncao grass has higher active components than those of forests, which can fundamentally solve the "contradiction between fungi and forests" and achieve sustainable development.

Third, replacing grain with Juncao grass to develop livestock industry. In 1990, the author started the research on using Juncao grass to cultivate shiitake mushroom, and the spent substrate packs were used to replace part of the concentrate feed to raise pigs. Since then, for more than 20 years, a series of achievements have been made in the research of Juncao forage, substrate feed, spent substrate feed, spent substrate extract of edible and medicinal fungi as feed additives, establishing the circular production and industrial development model involving plants, mushrooms and animals. The spent fungi substrate packs of *Ganoderma lucidum* cultivated by Juncao grass used as feed can improve the immunity of dairy cows.

Fourth, the development of biomass energy. In June 2008, joint experiments were carried out with Lanxi thermal power plant in Zhejiang Province to use Juncao grass for power generation. The measured result showed that the combustion power generation of Giant Juncao grass per hectare is equivalent to that of 52.5~60 tons of raw coal.

Fifth, "replacing wood with grass" to develop biological material. The research started in 2009, including using Giant Juncao and Lvzhou No.1 grassesto produce fiberboard, make pulp and paper, and produce activated carbon. The two species are new high-quality biological materials for industrial development. Due to its high yield and wide adaptability, Juncao used as materials has a huge potential and bright prospect.

Sixth, Juncao technology is used for ecological management. Since 1993, the technology has been used in Fujian, Chongqing, Yunnan, Xinjiang, Tibet, Guizhou and other provinces (autonomous regions and municipalities), as well as nine provinces along the Yellow River for ecological management. Relevant researches and experiments have been conducted for the treatment of soil erosion, collapsing hills, desertification, Pisha sandstone, proluvial fan, river bank and flood land, and saline-alkali land. We have overcome the key technical problems of Juncao ecological barrier construction and industrial development in ecologically fragile areas along the Yellow River. From 2011 to 2021, the experiments of planting Giant Juncao grass were conducted to control soil erosion in China-aided Agricultural Technology Demonstration Center in Robona, Rwanda, the source of the Nile River in Africa. On 30 October, 2012, with a rainfall of 51.4 mm in 3 hours, all the rainfall was intercepted in the experimental area by the Giant Juncao grass. Compared with the traditional crops such as corn, Giant Juncao grass reduces the soil loss by more than 97%. It is simple and easy to intercrop Juncao grass with dry crops such as corn, sweet potato and soybean, which yields faster results and bigger economic returns and call for smaller investment. We have made a series of achievements in the application of Juncao grass for ecological management at home and from abroad, developed a technical system and blazed a new trail for ecological restoration.

Promotion and application of Juncao technology

In May 1987, Juncao technology was applied in poverty-stricken areas in the northwest of Fujian Province, China. After obtaining good results, it was listed by Fujian Province as a project of using science and technology to prosper agriculture, performing actual deeds for the people, providing pairing assistance to Ningxia Province, intellectual assistance to Xinjiang Uygur Autonomous Region and Tibet Autonomous Region. Juncao project also was listed as the top priority project of National Spark Program by State Scientific and Technological Commission of China and as the preferred project among other technical projects by China Foundation for Poverty Alleviation. So far, Juncao technology has been promoted and adopted in 31 provinces (autonomous regions and municipalities) and 506 counties in China, making positive contributions to poverty alleviation.

Since 1994, Juncao technology has been listed by the Chinese government as an aid project for developing countries, and listed by UNDP as the priority cooperation project between China and other developing countries. In May 2017, Juncao technology was listed by the United Nations as a key project of the China-UN Peace and Development Fund.

Juncao technology facilitate the implementation of the 2030 Agenda for 13 Sustainable Developments Goals (SDGs) including: No Poverty, Zero Hunger, Good Health and Well-being, Quality Education, Gender Equality, Affordable and Clean Energy, Decent Work and Economic Growth, Industry, Innovation and Infrastructure, Reduced Inequalities, Responsible Consumption and Production, Climate Action, Life on Land, and Partnerships for the Goals. Through China aid and international cooperation, Juncao technology has been spread to 106 countries, and the demonstration centers (bases) have been established in Papua New Guinea, Fiji, Lesotho, Rwanda, Central African Republic, South Africa, Madagascar, Nigeria, Namibia, Malaysia, Myanmar, The Lao People's Democratic Republic, Thailand, North Korea, Brazil, providing China's solutions for international poverty reduction and the implementation of the United Nations 2030 Agenda for Sustainable Development.

On 2 September, 2021, the China International Development Cooperation Agency and Foreign Affairs Office of the Fujian Provincial People's Government co-organized the Forum on the 20th Anniversary of Juncao Assistance and Sustainable Development Cooperation in Beijing. António Guterres, secretary-general of the UN, sent a congratulatory letter to the Forum. H.E. James Marape, Prime Minister of the Independent State of Papua New Guinea, H.E. Faustin-Archange Touadéra, President of the Central African Republic, H.E. Imran Khan, Prime Minister of Pakistan, H.E. Phankham Viphavanh, Prime Minister of the Laos People's Democratic Republic, and H.E. Henry Puna, Secretary General of the Pacific Island Forum sent congratulatory videos to the Forum. The Forum is a major push to advance the development of the Juncao cause globally. In the future, the wider application of Juncao technology around the world will change the agricultural production mode and industrial development mode to a certain extent, greatly improve the living standards of thousands of impoverished farmers, and plays an active role in building a community of shared future for mankind.

In order to meet the needs of the development of Juncao industry at home and abroad, we have compiled the main achievements in the research and application of Juncao technology into a book.

To help readers have a better understanding of Juncao technology and its application scenarios, this book consists of 355 pictures and both Chinese and English captions.

In the process of writing this book, we'd like to express our gratitude to all academicians and experts who give us careful and helpful guidance, and to all colleagues who give us kind support. Meanwhile, we would also like to express our heartfelt thanks to all leaders who support Juncao industry and the scientific and technological circles and producers engaged in the Juncao industry.

Lin Zhanxi, Lin Dongmei
March 2022

目 录
CONTENTS

第一章
菌草技术由来 1
Chapter 1　Origin of Juncao Technology

菌草与菌草技术的起源 3
Origin of Juncao and Juncao Technology

菌草、菌草技术、菌草业的定义 7
Definition of Juncao, Juncao Technology and Juncao Industry

第二章
菌草草种选育与栽培 9
Chapter 2　Juncao Species Screening and Breeding

菌草选育目标 11
Screening and Breeding Goals of Juncao Grass

菌草主要品种 12
Main Varieties of Juncao Grass

第三章
菌草生态治理 27
Chapter 3　Juncao Ecological Management

菌草治理水土流失 29
Soil Erosion Control with Juncao Grass

菌草防风固沙 ··· 35
Wind Prevention and Sand Fixation on the Coast with Juncao Grass

菌草治理盐碱地 ··· 37
Treatment of Saline-alkali Land with Juncao Grass

菌草用于矿山植被修复 ··· 38
Mine Vegetation Restoration with Juncao Grass

第四章
菌草综合利用 ··· 39
Chapter 4　Comprehensive Application of Juncao

菌草栽培食药用菌 ·· 42
Cultivation of Edible and Medicinal Fungi with Juncao Grass

菌草与菌草菌物饲料 ·· 54
Juncao and Juncao Fungi Forage

菌草生物肥料 ·· 59
Juncao Biological Fertilizer

菌草生物质能源 ·· 60
Juncao Biomass Energy

菌草生物质材料 ·· 61
Juncao Biomass Material

第五章
助力脱贫攻坚和乡村振兴 ·· 63
Chapter 5　Contribution to Poverty Alleviation and Rural Revitalization

菌草技术培训 ·· 66
Juncao Technology Training

菌草技术示范基地 ·· 71
Juncao Technology Demonstration Bases

全国菌草扶贫会议 ·· 82
The National Symposiums on Poverty Alleviation with Juncao Technology

菌草创新产业园 ······ 87
Juncao Innovation Industrial Park

第六章
服务国际减贫和可持续发展 ······ 93
Chapter 6　Promotion of International Poverty Reduction and Sustainable Develepment

国际交流合作 ······ 95
International Exchange and Cooperation

人才培养 ······ 107
Talent Fostering

中国援建菌草技术示范中心 ······ 125
Juncao Technology Demonstration Centers Established with Aid from China

中国—联合国和平与发展基金菌草技术重点项目 ······ 182
Juncao Technology Key Project of the China-United Nations Peace and Development Fund

第一章 菌草技术由来

Chapter 1
Origin of Juncao Technology

菌草与菌草技术的起源

Origin of Juncao and Juncao Technology

中国是人类食用、药用和人工栽培食药用菌历史最悠久的国家，目前世界上广泛进行人工栽培的香菇、黑木耳、灵芝等十多种食药用菌，大部分都起源于中国。

China has the longest history of consuming and artificially cultivating edible and medicinal fungi. At present, there are more than ten kinds of edible and medicinal fungi, such as *Lentinus edodes*, *Auricularia auricula* and *Ganoderma lucidum*, widely cultivated in the world, most of which originate from China.

20世纪70年代，中国的香菇、木耳、灵芝等食药用菌栽培以阔叶树为原料，大规模发展食药用菌的生产后，与森林生态平衡之间产生"菌林矛盾"。

In the 1970s, the traditional cultivation of edible and medicinal fungi such as *Lentinus edodes*, *Auricularia auricula* and *Ganoderma lucidum* used wood sawdust from broad-leaved trees as raw materials, thus causing a contradiction between the development of fungus production and ecological balance.

用于栽培食用菌的阔叶树木材
The broad-leave trees cut for cultivating edible fungi

中国福建各地野生资源丰富，1972年著者林占熺提出用芒萁、五节芒等野生草本植物替代段木栽培香菇，1983年为了帮助贫困农户脱贫和保护生态环境，开始开展"以草代木"栽培食用菌的研究，1986年获得成功，发明了用野生草本植物栽培食药用菌的新技术。此后又开始"种草栽菇"，1987年首次在福建连城科技实验站种植类芦，1989年在福建尤溪县尤溪河滩种植象草，1993年在福建长汀禾田乡重度水土流失地种植类芦和象草，把生态治理与发展食用菌生产结合起来。

There are abundant wild resources such as *Dicranopteris dichotoma* and *Miscanthus floidulus* in all parts of Fujian Province, China. In 1972, the author Lin Zhanxi proposed to use the above-mentioned wild herbs instead of wood to cultivate shiitake mushrooms. In 1983, to help smallholder farmers get rid of poverty and protect the ecological environment, the research on "replacing wood with grass" to cultivate edible fungi began. In 1986, he succeeded in his experiment and invented a new technology. Since then, he has started "cultivating mushrooms with grass". In 1987, he first planted *Neyraudia reynaudiana* at Science and Technology Experimental Station of Liancheng County, Fujian Province. In 1989, *Pennisetum purpureum* was planted in Youxi River Beach, Youxi County, Fujian Province. In 1993, *Neyraudia reynaudiana* and *Pennisetum purpureum* were planted in the heavily soil-eroded land in Hetian Township, Changting County, Fujian Province, which combined ecological management with the development of edible fungus production.

"以草代木"栽培食药用菌具有见效快、经济效益高、生态效益好、适应性广、千家万户能参与等特点，1991年被福建省科委和中国国家科委列为"八五"国家级星火计划重中之重项目，被中国扶贫基金会列为扶贫首选项目；1991~1995年在福建51个县实施，菌草栽培食用菌示范生产12.39亿筒（袋），产值增加22.46亿元，节约阔叶树51.26万立方米，增加就业49万人。

Cultivation of edible and medicinal fungi with grass instead of wood yields faster results, bigger economic returns, and good ecological benefit. Juncao grass has wide adaptability, and thousands of households can participate in the industry. In 1991, Juncao project was listed as the top priority project of the Eighth Five-Year Plan by the Scientific and Technological Commission of Fujian Province and China, and was listed as the first-choice project by the China Foundation for Poverty Alleviation. From 1991 to 1995, Juncao project was implemented in 51 counties in Fujian Province. The demonstration production of edible fungi cultivated with Juncao grass was 1.239 billion tubes (bags), the output value increased by 2.246 billion yuan, 512,600 cubic meters of broad-leaved trees were saved, and 490,000 jobs were created.

"以草代木"栽培食药用菌的成功应用，引起社会各界的广泛关注。为了适应农村脱贫和发展中国家发展菌草业的需要，1996年11月27~28日，中国国际经济交流中心与福建省科委共同举办了首届菌草业发展国际研讨会，确定了"菌草""菌草技术""菌草业"的科学定义。

The successful application of "substituting wood with grass" to cultivate edible and medicinal fungi has attracted widespread attention from all walks of life. In order to meet the needs of rural poverty alleviation and the development of Juncao industry in developing countries, on 27-28 November, 1996, the China Center for International Economic Exchange and the Scientific and Technological Commission of

国家菌草工程技术研究中心启动仪式
Launching Ceremony of China National Engineering Research Center of Juncao Technology

Fujian Province co-organized the first International Symposium on the Development of Juncao Industry, in which the scientific definitions of "Juncao", "Juncao technology" and "Juncao industry" have been proposed.

为推动菌草科学的研究和促进菌草新型产业的发展，2002 年 8 月，时任福建省省长习近平支持建立世界首个菌草科学实验室；2009 年 8 月，科技部批准福建农林大学菌草研究所为国际科技合作基地；2011 年 12 月 30 日，科技部批准依托福建农林大学组建国家菌草工程技术研究中心；2012 年 10 月 28 日，国家发改委批准依托福建农林大学组建菌草综合开发利用技术国家地方联合工程研究中心；2018 年 12 月 14 日，教育部批准福建农林大学成立菌草生态产业省部共建协同创新中心。

In order to promote the research of Juncao science and the development of new Juncao industry, in August 2002, Xi Jinping, then governor of Fujian Province, supported the establishment of the world's first Juncao Science Laboratory. In August 2009, the Ministry of Science and Technology approved FAFU Juncao Technology Research Center as the International Scientific and Technological Cooperation Demonstration Base; On 30 December, 2011, the Ministry of Science and Technology approved the set-up of the China National Engineering Research Center of Juncao Technology in FAFU. On October 28, 2012, the National Development and Reform Commission approved the founding of National and Local Joint Research Center for Juncao Technology Comprehensive Development and Utilization in FAFU. On December 14, 2018, the Ministry of Education approved the establishment of Collaborative Innovation Center by the Ministry and the Province on Juncao Ecological Industry.

30 多年来，在中国政府支持下，菌草技术的科学研究从"以草代木"发展菌业拓展至菌草选育、菌草生态治理、菌草饲料、菌草菌物饲料、菌草生物肥料、菌草生物质能源和菌草生物质材料等领域，开辟了"菌"与"草"交叉的科学研究与应用新领域。

Over the past 30 years, with the support of the Chinese government, the scientific research of Juncao Technology has expanded from the development of fungi industry by "replacing wood with grass" to the fields of Juncao breeding, ecological management, Juncao forage, spent substrate feed, Juncao biological fertilizer, Juncao biomass energy and Juncao biomass materials, thus opening up a new field of scientific research and application in fungi and grass industries.

菌草、菌草技术、菌草业的定义

Definition of Juncao, Juncao Technology and Juncao Industry

为了适应菌草事业发展的需要，1996年11月27～29日在福州召开的首届菌草业发展国际研讨会确定了"菌草""菌草技术""菌草业"的定义。

To meet the development needs of Juncao cause, the definitions of "Juncao", "Juncao technology" and "Juncao industry" were confirmed at The First International Symposium on the Development of Juncao Industry held in Fuzhou on 27-29 November, 1996.

菌草：可以作为栽培食用菌、药用菌培养基的草本植物。

Juncao: Herbaceous plants that can be used as the culture substrate for cultivation of edible and medicinal fungi.

菌草技术：利用菌草作为培养基（基质）或原料，通过真菌的分解、促进和共生，生产菌物、功能性食品、饲料、肥料、生物质能源、生物材料，并用于环境保护的技术。

Juncao Technology: Techniques that utilize Juncao as medium substrate or raw material, through the decomposition, promotion and symbiosis of fungi, to produce mushrooms, functional food, feed, fertilizer, biomass energy, biological materials, and to apply in environmental protection.

菌草业：运用菌草、菌草技术及相关技术形成的可持续产业。

Juncao Industry: A sustainable industry formed by application of Juncao technology and other interrelated techniques.

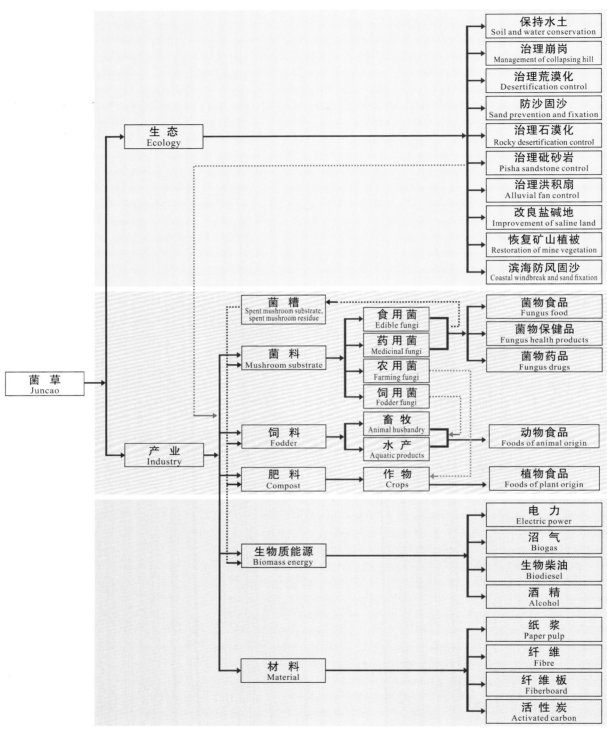

菌草技术与菌草产业体系示意图
Map of Juncao Technology and Juncao Industrial System

第二章 菌草草种选育与栽培

Chapter 2
Juncao Species Screening and Breeding

菌草选育目标

Screening and Breeding Goals of Juncao Grass

根据菌草产业发展需要提出菌草品种选育六个目标：

According to the development needs of Juncao industry, six goals for the screening and breeding of Juncao grass varieties are proposed:

（1）光合作用率高，可高效利用太阳能。

High photosynthesis rate enables efficiently utilization of solar energy.

（2）富含内生菌且固氮作用强。

Rich in endophytes with strong nitrogen-fixation.

（3）根系发达，蓄水固沙、固土、改土效果好。

Thriving root system with excellent effects in water storage and sand fixation, soil fixation and soil improvement.

（4）生长速度快，植株高大，生物量大。

Fast growth, tall plant, and large biomass.

（5）粗蛋白含量高，营养丰富，综合利用价值高。

Rich nutrition with high crude protein content and high comprehensive utilization value.

（6）抗逆性强，适应性广。

Strong in stress tolerance and wide adaptability.

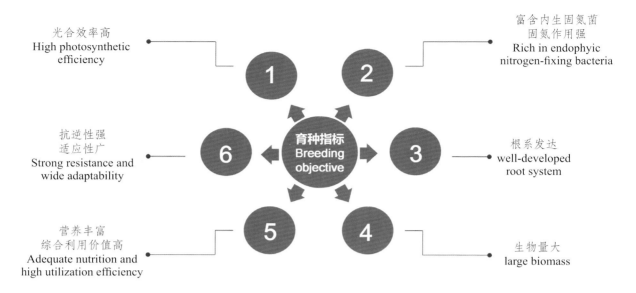

菌草品种选育六个目标
Six goals for the screening and breeding of Juncao grass varieties

菌草主要品种

Main Varieties of Juncao Grass

表 2-1　适宜栽培食药用菌的菌草
Table 2-1　Juncao suitable for cultivating edible and medicinal fungi

	门 Phylum	纲 Class	科 Family	属 Genus	种 Species
人工栽培 Artificial planting	被子植物门 Angiospermae	单子叶植物纲 Monocotyledoneae	禾本科 Gramineae	蒺藜草属 Cenchrus	巨菌草（*Cenchrus fungigraminus*）
					象草（*Cenchrus purpureum*）
					王草（*Cenchrus sinese*）
					杂交狼尾草（*Cenchrus americanum* × *C. purpureum*）
					紫色象草（*Cenchrus purpureum* cv. Purple）

第二章　菌草草种选育与栽培
Chapter 2　Juncao Species Screening and Breeding

续表

门 Phylum	纲 Class	科 Family	属 Genus	种 Species
人工栽培 Artificial planting	被子植物门 Angiospermae	单子叶植物纲 Monocotyledoneae	禾本科 Gramineae	甘蔗属 Saccharum	彼特草（*Saccharum robustum*）
				菅属 Themeda	菅（*Themeda gigantea* var. *villosa*）
				雀稗属 Paspalum	宽叶雀稗（*Paspalum wettsteinii*）
					毛花雀稗（*Paspalum dilatatum*）
				香根草属 Vetiveria	香根草（*Vetiveria zizanioides*）
				高粱属 Sorghum	拟高粱（*Sorghum propinquum*）
					苏丹草（*Sorghum sudanense*）
				香茅属 Cymbopogon	香茅（*Cymbopogon citratus*）
				芦竹属 Arundo	绿洲1号、2号、3号、5号、6号、9号、10号（*Arundo donax* cv. Lvzhou）
					莱竹（*Arundo donax*）
				大米草属 Spartina	大米草（*Spartina anglica* Hubbard）
				狗尾草属 Setaria	卡松古鲁狗尾草（*Setaria anceps*）
				摩擦禾属 Tripsacum	危地马拉草（*Tripsacum Laxum*）
			雨久花科 Pontederiaceae	凤眼莲属 Eichhornia	凤眼莲（*Eichhornia crassipes*）
		双子叶植物纲 Dicotyledoneae	豆科 Leguminosae	苜蓿属 Medicago	紫花苜蓿（*Medicago sativa*）
				柱花属 Stylosanthes	柱花草（*Stylosanthes guianensis*）
			菊科 Asteraceae	松香草属 Silphium	串叶草（*Silphium perfoliatum*）
			伞形科 Umbelliferae	阿魏属 Ferula	阿魏（*Ferula sinkiangensis*）

续表

	门 Phylum	纲 Class	科 Family	属 Genus	种 Species
野生 Wild	被子植物门 Angiospermae	单子叶植物纲 Monocotyledoneae	禾本科 Gramineae	芦苇属 Phragmites	芦苇（*Phragmites communis*）
				类芦属 Neyraudia	类芦（*Neyraudia reynaudiana*）
				芒属 Miscanthus	五节芒（*Miscanthus floridulus*）
					芒（*Miscanthus sinensis*）
				荻属 Triarrhena	荻（*Triarrhena sacchariflora*）
				甘蔗属 Saccharum	斑茅（*Saccharum arundinaceum*）
				菅属 Themeda	苞子草（*Themeda gigantea* var. *caudata*）
				芦竹属 Arundo	芦竹（*Arundo donax*）
				芨芨草属 Achnatherum	芨芨草（*Achnatherum splendens*）
				野古草属 Arundinella	野古草（*Arundinella hirta*）
					石芒草（*Arundinella nepalensis*）
	蕨类植物门 Pteridophyta	薄囊蕨纲 Lepidoptera	里白科 gleicheniaceae	芒萁属 Dicranopteris	芒萁（*Dicranopteris dichotoma*）
					大芒萁（*Dicranopteris ampla*）
作物秸秆 Crop straws	被子植物门 Angiospermae	单子叶植物纲 Monocotyledoneae	禾本科 Gramineae	甘蔗属 Saccharum	甘蔗（*Saccharum sinensis*）
				玉蜀黍属 Zea	玉米（*Zea mays*）
				小麦属 Triticum	小麦（*Triticum aestivum*）
				稻属 Oryza	稻（*Oryza sativa*）
			芭蕉科 Musaceae	芭蕉属 Musa	香蕉（*Musa nana*）

续表

续表

门 Phylum	纲 Class	科 Family	属 Genus	种 Species
作物秸秆 Crop straws	被子植物门 Angiospermae	双子叶植物纲 Dicotyledoneae	锦葵科 Malvaceae	棉属 Gossypium	棉花（Gossypium hirsutum）
			菊科 Asteraceae	向日葵属 Helianthus	向日葵（Helianthus annuus）

1. 巨菌草
Cenchrus fungigraminus

生物学特性： 直立丛生，植株高 3~6 米，人工栽培高可达 8.78 米。根系发达；茎秆粗壮，达 3.5 厘米，每个节均可长腋芽，可利用腋芽进行无性繁殖。在自然条件下，巨菌草一般不结籽，少数地区结籽，但发芽率低。

在热带、亚热带地区为多年生植物，温度 12℃时开始生长，适宜生长温度 25~35℃。

巨菌草是典型的 C4 植物，光合效率高，耐低、中度盐碱，适宜在年降水量 500 毫米以上地区生长。

Biological characteristics: Upright cluster, plant height of 3-6m, artificial culture up to 8.78m, developed root system, culms robust, diameter, up to 3.5cm, each node can grow axillary buds which can be used for asexual reproduction. Under natural conditions, The Giant Juncao grass does not produce seed, even in a few areas do produce seed, but germination rate is low.

巨菌草（Giant Juncao grass）
Cenchrus fungigraminus

It is a perennial plant in tropical and subtropical areas. It starts to grow at 12℃ and the optimal growth temperature is 25-35 ℃.

Giant Juncao grass is a typical C4 plant with high photosynthetic efficiency, resistance to low and moderate salinity and alkalinity. It is suitable to grow in areas with annual precipitation of more than 500mm.

用途： 用于保持水土，栽培香菇、毛木耳、黑木耳、平菇、双孢蘑菇、灵芝、猴头菌等50余种食药用菌。作牛、羊、马、猪、鹿、鸡、鸭、鹅等家畜家禽的饲料，还可用于生产板材、制浆造纸、燃烧发电、提取酒精、生产沼气等。

Application: The plant is for the wide usage of water and soil conservation, cultivation of more than 50 kinds of edible and medicinal fungi such as *Lentinula edodes*, *Auricularia polytricha*, *Auricularia auricula*, *Pleurotus ostreatus*, *Agaricus bisporus*, *Ganoderma lucidum* and *Hericium erinaceus*, production of feed for cattle, sheep, horses, pigs, deer, chickens, ducks, geese and other livestock and poultry, production of fiberboard, pulp and paper, power generation, extraction of alcohol, production of biogas, etc.

2. 象草
Cenchrus purpureum

生物学特性： 丛生，植株高3~5米，根系发达。在热带、亚热带地区为多年生，抽穗结籽，但种子发芽率低。1983年在福建农学院菌草圃种植的象草，每年收割1~2次，已连续生长38年仍正常生长。

Biological characteristics: Standing upright in clumps, the plant can be up to 3-5m high with well-developed root system. It is perennial in tropical and subtropical areas. It bears seeds during heading, but the seed germination rate is low. This elephant grass planted in the nursery of Fujian Agricultural College in 1983 harvested once or twice a year, has been growing for 38 consecutive years and still grows normally.

用途： 用于保持水土，栽培香菇、毛木耳、黑木耳、平菇、双孢蘑菇、灵芝、猴头菌等50余种食药用菌。作牛、羊、马、猪、鹿、鸡、鸭、鹅等家畜家禽的饲料，还可用于生产板材、纸浆、燃烧发电、生产沼气等。

象草（Elephant Grass）
Cenchrus purpureum

Application: The plant is for the wide usage of water and soil conservation, cultivation of more than 50 kinds of edible and medicinal fungi such as *Lentinula edodes*, *Auricularia polytricha*, *Auricularia auricula*, *Pleurotus ostreatus*, *Agaricus bisporus*, *Ganoderma lucidum* and *Hericium erinaceus* production of feed for cattle, sheep, horses, pigs, deer, chickens, ducks, geese and other livestock and poultry, production of fiberboard, pulp and paper, power generation, extraction of alcohol, production of biogas, etc.

3. 菅
Themeda gigantea var. *villosa*

生物学特性： 在热带、亚热带地区多年生。丛生，植株高达3米以上，结籽，可播种繁育。耐旱、耐瘠薄。

Biological characteristics: It is perennial in tropical and subtropical areas. Standing upright in clumps, the plant can be up to 3m high, produces seeds that can be bred and has resistance to drought and barren land.

用途： 可用于保持水土，还可栽培香菇等50余种食药用菌。栽培食药用菌后的菌糟可作猪、牛、羊等牲畜的饲料。

Application: The plant is for the use of water and soil conservation, cultivation of more than 50 kinds of edible and medicinal fungi such as *Lentinula edodes*. The spent mushroom substrate can be used as feed for pigs, cattle, sheep and other livestock.

菅（Villous Themeda）
Themeda gigantea var. *villosa*

4. 大米草
Spartina anglica Hubbard

生物学特性： 多年生。具根状茎。秆高150~200厘米。耐盐碱，生长在滩涂。在福建罗源湾，一株大米草1年可扩至9平方米，每公顷年产鲜草15吨。

Biological characteristics: Perennial. The plant has rhizome. Culms 150-200cm tall. It is saline alkali resistant and grows on the beach. In Luoyuan Gulf of Fujian Province, one stock of *Spartina anglica* can expands to $9m^2$ in just one year. The annual yield of fresh grass is 15 tons per hectare.

用途： 固滩护堤、促淤造陆；作畜禽饲料；可与巨菌草等其他菌草混合栽培香菇、毛木耳、黑木耳、平菇、竹荪、猴头菌等多种食药用菌。

Application: The plant is for the wide usage of consolidating beach, protecting dam, promoting silting to create land, and for livestock and poultry feed. It can be mixed with Giant Juncao grass and other grasses to cultivate some varieties of edible and medicinal fungi such as *Lentinula edodes*, *Auricularia auricula*, *Pleurotus ostreatus*, *Dictyophora* spp., *Hericium erinaceus* and so on.

大米草（Common Cordgrass）
Spartina anglica Hubbard

5. 宽叶雀稗
Paspalum Wettsteinii Hackel

生物学特性： 在热带、亚热带低海拔地区为多年生。丛生，株高160厘米左右。

Biological characteristics: It is perennial in tropical and subtropical low altitude areas. Standing upright in clumps, the plant can be up to 160cm high.

用途： 可作饲料及栽培香菇、木耳、金针菇、平菇、灵芝等多种食药用菌的培养料。

宽叶雀稗（Broadleaf Dallisgrass）
Paspalum Wettsteinii Hackel

Application: It can be used as feed for livestock and culture material for various edible and medicinal fungi such as *Lentinula edodes*, *Auricularia auricula*, *Flammulina velutipes*, *Pleurotus ostreatus* and *Ganoderma lucidum*.

6. 香根草
Vetiveria zizaniodes （L.）Nash

生物学特性： 多年生，根具有特殊的芳香味。秆高 1~2.5 米，直立，密丛生，在 -15℃以上条件下可越冬。具有耐旱、耐湿、耐瘠薄、耐寒的特点。

Biological characteristics: Perennial, the root has a special fragrance. Culms 1-2.5m tall, erect in dense clumps. It can survive the winter at -15℃, which has strong resistance to drought, moisture, infertility and coldness.

用途： 保持水土，用作香菇、毛木耳、黑木耳、平菇、竹荪、猴头菌、灵芝等食药用菌的培养料。

Application: The plant is for the use of water and soil conservation and culture material for edible and medicinal fungi such as *Lentinula edodes*, *Auricularia polytricha*, *Auricularia auricula*, *Pleurotus ostreatus*, *Dictyophora* spp., *Hericium erinaceus* and *Ganoderma lucidum*.

香根草（Vetiver）
Vetiveria zizaniodes (L.) Nash

7. 香茅
Cymbopogon citratus （DC.） Stapf

生物学特性：多年生，丛生，高达2米，秆粗壮，有柠檬香气。

Biological characteristics: Perennial, cluster, up to 2m high, culms robust, with lemon aroma.

用途：保持水土，与禾本科、豆科菌草混合可用于栽培多种食药用菌。

Application: For water and soil conservation. It can be mixed with gramineous and leguminous Juncao grass to cultivate varieties of edible and medicinal fungi.

香茅（Lemongrass）
Cymbopogon citratus （DC.） Stapf

8. 拟高粱
Sorghum propinquum （Kunth） Hitchc

生物学特性：秆高2~3米，叶宽达3厘米，长达90厘米。须根发达。结籽，用种子播种繁育。耐高温、耐旱、耐酸，可在pH 4~5.5的酸性土壤种植。

Biological characteristics: Culms 2-3m tall, the leaf width is 3cm and the length 90cm. Fibrous roots are developed. The plant produces seeds that can be bred. With resistance to high temperature, drought and acidity, it can be planted in acid soil with a pH of 4-5.5.

用途：作为栽培香菇、金针菇、平菇、猴头菌、灵芝等食药用菌的培养料；栽培食药用菌的菌糟可作饲料、菌料和肥料。

Application: It can be used as the culture material of *Lentinula edodes*, *Flammulina velutipes*, *Pleurotus ostreatus*, *Hericium erinaceus*, *Ganoderma lucidum* and other edible and medicinal fungi. The substrate of cultivated edible and medicinal fungi can be used as feed, fungi material and fertilizer.

拟高粱（Pseudo-sorghum）
Sorghum propinquum （Kunth） Hitchc

9. 绿洲 1 号
Arundo donax L.

生物学特性： 多年生，秆粗壮，高 2~6 米，人工种植可高达 9.5 米。具粗而多节的根状茎，形似竹笋，耐瘠薄、耐低温，在 -23℃条件下能安全越冬。

Biological characteristics: Perennial. Culms robust, 2-6m high, the height can be up to 9.5m under artificial cultivation. It has thick rhizomes with many nodes like bamboo. It is able to bear poor environment and low temperatures, which means that it can overwinter safely even when the temperature is as low as -23℃.

用途： 生态治理，是荒漠化地区恢复植被的优良草种；作栽培香菇、木耳、灵芝等 50 余种食药用菌的培养料；用于生产纸浆、板材等。

Application: For ecological control, it is an excellent grass species for vegetation restoration in desertification areas. It is used as culture material for more than 50 kinds of edible and medicinal fungi such as *Lentinula edodes*, *Auricularia auricula* and *Ganoderma lucidum*; production of pulp, fibre board, etc..

绿洲 1 号（Lvzhou No.1）
Arundo donax L.

10. 串叶草
Silphium perfoliatum L.

生物学特性： 多年生。茎直立，茎秆四棱形，株高 2~3 米，根粗壮，-37℃能越冬。

Biological characteristics: Perennial. The culm is upright and quadrangular, the plant height is 2-3m, with thick roots, and can survive the winter at -37℃.

用途： 作为饲料饲养家禽家畜。与禾本科的菌草混合作平菇、阿魏菇、杏鲍菇等多种食用菌的培养料。

Application: Feed for poultry and livestock. Mixed with gramineous Juncao grass as culture material for *Pleurotus ostreatus*, *Pleurotus ferulae*, *Pleurotus eryngii* and other edible fungi.

串叶草（Cup plant）
Silphium perfoliatum L.

11. 芒萁

Dicranopteris dichotoma（Thunb）Bernh

生物学特性： 多年生，株高40~120厘米。在巴布亚新几内亚东高省的热带雨林中发现高达14.7米、茎粗达0.6厘米、株重507克的芒萁。耐旱、耐酸、耐瘠薄，是酸性土壤指示植物。

Biological characteristics: Perennial, 40-120cm high. A plant that was 14.7m in height, 0.6cm in stem diameter and 507g in weight, was found in the tropical rain forest of Eastern Highland Province of PNG. It is tolerant to drought, acidity and infertility, which is an indicator plant of acid soil.

用途： 与五节芒、类芦、斑茅、芦苇、巨菌草、象草、芦竹等菌草配合作栽培香菇、黑木耳、毛木耳、金针菇、平菇、灵芝、猴头菌等多种食药用菌的培养料。

Application: It can be applied for the cultivation of *Lentinula edodes*, *Auricularia auricula*, *Auricularia polytricha*, *Flammulina velutipes*, *Pleurotus ostreatus*, *Ganoderma lucidum* and *Hericium erinaceus* together with the other Juncao grass species, such as *Miscanthus floridulus*, *Neyraudia reynaudiana*, *Saccharum arundinaceum*, *Phragmites communis*, *Pennisetum purpureum*, *Arundo donax*, etc.

芒萁（Mang Qi）
Dicranopteris dichotoma（Thunb）Bernh

12. 类芦
Neyraudia reynaudiana(kunth) Keng

生物学特性： 在中国南方多年生。秆高 1~4.8 米，人工栽培高可达 5.2 米以上。具木质根状茎。类芦耐旱、耐瘠薄。

Biological characteristics: Perennial in the south of China. Culms 1-4.8m tall, while artificially cultivated it can reach up to more than 5.2m. With woody rhizome, the plant has resistance to drought and barren soil.

用途： 保持水土；幼嫩时可作牛、马饲料；用于栽培香菇、木耳、金针菇、平菇、灵芝、猴头菌、竹荪等 50 余种食药用菌；其栽培食药用菌后的菌糟可作饲料、菌料和肥料。

Application: water and soil conservation. The fresh and tender plant is high-quality feed for cows and horses. It can be used to cultivate more than 50 kinds of edible and medicinal fungi such as *Lentinula edodes*, *Auricularia* spp., *Flammulina velutipes*, *Pleurotus ostreatus*, *Ganoderma lucidum*, *Hericium erinaceus*, *Dictyophora* spp. The spent mushroom substrate packs can be used as livestock feed, fungus material and fertilizer.

类芦（Burma reed）
Neyraudia reynaudiana(kunth) Keng

13. 斑茅
Saccharum arundinaceum Retz.

生物学特性: 多年生,丛生。高2~4.5米,秆粗壮,茎直径达2厘米。

Biological characteristics: Perennial, forming large clumps. Culms robust, 2-4.5m tall, 2cm in diameter.

用途: 幼嫩时可作牛、马的饲料,成熟后可用于造纸。亦可作香菇、木耳、金针菇、灵芝、猴头菌、竹荪等50余种食药用菌的培养料。

Application: It can be used as fodder for cows and horses when young and tender. When getting mature, it can be used to make paper, as well as the substrate material for cultivation of more than 50 kinds of edible and medicinal fungi species, such as *Lentinula edodes*, *Auricularia* spp., *Flammulina velutipes*, *Ganoderma lucidum*, *Hericium erinaceus* and *Dictyophora* spp..

斑茅（Reedlike Sugarcane）
Saccharum arundinaceum Retz.

14. 芦苇
Phragmites communis Trin

生物学特性: 多年生,具粗壮根状茎。茎秆中空似竹,高1~4米。对温度适应性强,在-32℃条件下能越冬。

Biological characteristics: Perennial, robust rhizomes. Culm is cannular and 1-4m tall. It has strong adaptability to temperature and can survive winter at -32℃.

用途: 用于护岸固堤;嫩时可用作饲料;秆供编席、织帘、造纸;作为香菇、毛木耳、黑木耳、盾形木耳、灵芝、竹荪等50余种食药用菌的培养料;菌糟亦可作饲料。

Application: For revetment and embankment consolidation. It can be used as feed when young and

芦苇（Reed）
Phragmites communis Trin

tender. Culms can be used for weaving mat, curtain and paper-making. It can also be used as the substrate material for cultivation of more than 50 kinds of fungi species, such as *Lentinula edodes*, *Auricularia polytricha*, *Auricularia auricula*, *Auricularia peltata*, *Ganoderma lucidum* and *Dictyophora* spp.. Furthermore, the spent substrate can also be utilized as the fungi forage.

15. 五节芒
Miscanthus floridulus（Labill.）Warb.

生物学特性： 多年生，丛生，秆高 2~4 米。
Biological characteristics: Perennial and cluster. Culms 2-4m tall.

用途： 幼嫩时可作牛饲料；茎既可用于造纸，又可用作栽培香菇、毛木耳、黑木耳、灵芝、竹荪、猴头菌、平菇、金针菇、滑菇、朴菇等50余种食药用菌的培养料；菌糟可作猪、牛、羊等牲畜的饲料。
Application: It can be used as the fodder for cows when young and tender. The stem can be utilized to make paper. It can also be applied as the substrate for cultivation of more than 50 kinds of edible and medicinal fungi species, such as *Lentinula edodes*, *Auricularia polytricha*, *Auricularia auricula*, *Ganoderma lucidum*, *Dictyophora* spp., *Hericium erinaceus*, *Pleurotus ostreatus*, *Flammulina velutipes*, *Pholiota nameko* and *Agrocybe aegerita*. The spent substrate after cultivation can also be applied as fodder for livestock like pigs, cows, sheep, etc..

五节芒（Fivenodes Awngrass）
Miscanthus floridulus（Labill.）Warb.

16. 紫色象草
Cenchrus purpureum cv. Purple

生物学特性：形态特征与象草相似，茎叶紫色，故称紫色象草。紫色象草喜暖湿气候，适应性广，在海拔1000米以下、年降水量500毫米以上的热带、亚热带地区均可种植。适应性强，生长快，多年生，产量高，且耐肥、耐旱、耐酸，抗倒伏性强。种植1次可连续收割15年以上。

Biological characteristics: The morphological characters are similar to those of elephant grass. Its stems and leaves are purple, so it is called purple elephant grass. It likes warm and humid climate and has a wide adaptability. It can be planted in tropical and subtropical areas with an altitude of less than 1000m and an annual precipitation above 500mm. It has strong adaptability, fast growth, perennial, high yield, and is resistant to fertilizer, drought, acidity and lodging. Planting once can harvest continuously for more than 15 years.

用途：水土保持；作食药用菌栽培的培养料；作家禽家畜饲料。

Application: For water and soil conservation. It can be used as culture material for edible and medicinal fungi and feed for livestock and poultry.

紫色象草（Purple Elephant Grass）
Cenchrus purpureum cv. Purple

第三章 菌草生态治理

Chapter 3
Juncao Ecological
Management

菌草用于治理水土流失、崩岗、荒漠化、石漠化、砒砂岩、洪积扇、盐碱地，以及防风固沙、矿山植物修复、治理土壤重金属污染等。菌草用于生态治理效果好、见效快、投入省，开辟了一条生态效益与经济效益、社会效益相结合的可持续发展的新途径。

Juncao grass is used to control soil erosion, hill collapse, desertification, rocky desertification, arsenic sandstone, alluvial fan, saline-alkali land, as well as for wind prevention and sand fixation, restoration of mine vegetation, and treatment of heavy metal pollution in soil. Juncao grass used for ecological management has good effects, quick results, and low investment, opening up a new way of sustainable development that integrates ecological benefits with economic benefits and social benefits.

菌草治理水土流失

Soil Erosion Control with Juncao Grass

1. 种植菌草治理水土流失
Soil Erosion Control with Juncao Grass

在福建坡度为 45° 的水土流失地等高线种植菌草，治理当年见效
Juncao grass planted on the contour line of the soil erosion area with a slope of 45 degrees takes effect on the same year in Fujian

2. 菌草治理崩岗
Collapsing Hill Treatment with Juncao Grass

巨菌草根系发达，用于治理崩岗，3个月见效

With a well-developed root system, Giant Juncao grass is used to treat collapsing hills, which takes effect within three months

3. 菌草治理荒漠化
Desertification Control with Juncao Grass

在宁夏戈壁滩地种植菌草，每公顷产鲜草300吨
Juncao grass planted in the Gobi Desert in Ningxia Hui Autonomous Region produces 300 tons of fresh grass per hectare

4. 菌草治理石漠化
Rocky Desertification Control with Juncao Grass

在贵州、广西等地种植菌草治理石漠化并发展畜业
Juncao grass planted in Guizhou Province and Guangxi Zhuang Autonomous Region controls rocky desertification and develops animal husbandry

5. 菌草治理砒砂岩
Arsenic Sandstone Treatment with Juncao Grass

在内蒙古鄂尔多斯种植菌草治理砒砂岩
Juncao grass is planted in Ordos, Inner Mongolia Autonomous Region to treat arsenic sandstone

6. 菌草治理洪积扇
Proluvial Fan Control with Juncao Grass

2015~2020 年，在青海省贵德海拔 2353 米处种植菌草治理洪积扇，每公顷产巨菌草鲜草 135 吨
Juncao grass is planted to control proluvial fans in Guide, Qinghai Province from 2015 to 2020 at an altitude of 2353m. The production of Giant Juncao grass is 135 tons per hectare

菌草防风固沙

Wind Prevention and Sand Fixation on the Coast with Juncao Grass

1. 菌草治理流动沙丘
Mobile Sand Dunes Control with Juncao Grass

2011年7月，在西藏林芝海拔3067米处种植菌草治理流动沙丘，当年即把流动沙丘固住。绿洲1号在当地可多年生

In July 2011, Juncao grass was planted in the mobile sand dunes at an altitude of 3067m in Linzhi, Tibet Autonomous Region. The mobile sand dunes were fixed in that year, and Lvzhou No. 1 was perennial there

2. 菌草治理流动、半流动沙地
Treatment of Mobile and Semi-mobile Sandy Land with Juncao Grass

在内蒙古乌兰布和沙漠流动、半流动沙地种植菌草固沙，生长115天，每丛可固沙18.8立方米。2013年种植的巨菌草到2021年仍具固沙作用

Juncao grass is planted in the mobile and semi-mobile sandy land of the Ulan Buh Desert in Inner Mongolia Autonomous Region to fix sand. It grows for 115 days and the clumps can fix $18.8m^3$ of sand. The Giant Juncao grass planted in 2013 can still fix sand in 2021

2018年，在内蒙古乌兰布和流动沙漠地种植巨菌草，每公顷产鲜草183吨，每公顷根重162吨

In 2018, Giant Juncao grass was planted in the mobile desert of Ulan Buh Desert, Inner Mongolia Autonomous Region, produces fresh grass 183 tons per hectare and the roots weight 162 tons per hectare

3. 滨海菌草防风固沙
Wind Prevention and Sand Fixation on the Coast with Juncao Grass

在福建平潭滨海种植绿洲 1 号和巨菌草后 2 个月即把流动沙地固住
Lvzhou No. 1 and Giant Juncao grass were planted in the coastal area of Pingtan, Fujian Province, and the mobile sandy land was fixed in two months after planting

菌草治理盐碱地

Treatment of Saline-alkali Land with Juncao Grass

2018~2021 年，在福建平潭种植菌草治理改良盐碱地获得成功
From 2018 to 2021, Juncao grass planted in Pingtan, Fujian Province to improve saline-alkali land met with great success

菌草用于矿山植被修复

Mine Vegetation Restoration with Juncao Grass

在福建泉州种植菌草使矿山复绿
Treatment of Quanzhou Mine in Fujian Province with Juncao grass

在福建罗源种植菌草治理石矿山
Treatment of Luoyuan Stone Mine in Fujian Province with Juncao grass

第四章
菌草综合利用

Chapter 4
Comprehensive Application of Juncao

第四章 菌草综合利用
Chapter 4　Comprehensive Application of Juncao

　　菌草是新型的生物材料、生物质能源和农业资源。用于生态治理具有效果好、投入省、见效快的优点，可用于"以草代木"栽培食药用菌，"以草代粮"作饲料发展畜业，"以草代煤"用于发电，还可"以草代木"用于造纸、制板材等，实现"一草多用""综合利用"。菌草产业是高产、优质、高效、生态、安全的新型生态产业，具有产业链长、覆盖面广、潜力巨大、应用前景广阔的优势。

　　Juncao grass is a new type of biological material, biomass energy and agricultural resources, which has the advantages of good effect, low investment and quick result when used for ecological management. Juncao grass replacing wood can be used to cultivate edible and medicinal fungi, Juncao grass replacing grain as feed to develop animal industry, Juncao grass replacing coal for power generation, and Juncao grass replacing wood for papermaking and board making. The Juncao industry is a high-yield, high-quality, high-efficiency, ecological, and safe industry. It has a long industrial chain, wide coverage, huge potential and broad application prospects.

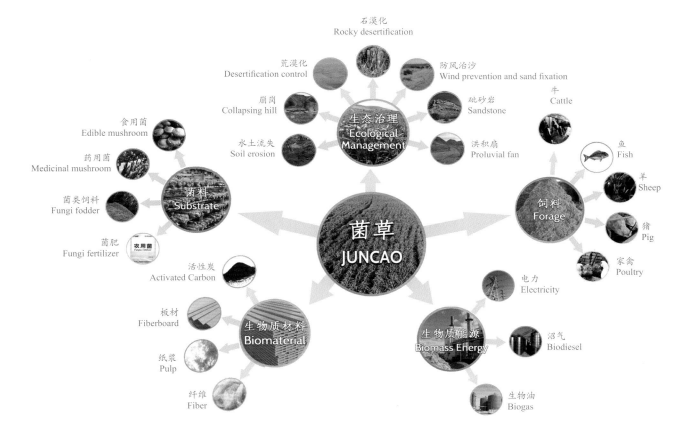

菌草产业示意图

Diagram of Juncao Industry

菌草栽培食药用菌

Cultivation of Edible and Medicinal Fungi with Juncao Grass

1. 菌草栽培食药用菌种类
Varieties of Edible and Medicinal Fungi Cultivated with Juncao Grass

1983~2020年，已筛选、培育适用菌草栽培的食药用菌15个科55个品种。

From 1983 to 2020, 55 species of edible and medicinal fungi from 15 families cultivated with Juncao grass have been screened and bred.

表 4-1　适宜用菌草栽培的食药用菌种类

Table 4-1　Types of edible and medicinal fungi suitable for the cultivation of Juncao grass

门 Phylum	纲 Class	亚纲 Subclass	目 Order	科 Family	属 Genus	种 Species
担子菌亚门 Basidiomycotina	层菌纲 Hymenomycetes	同隔担子菌亚纲 Holobasidiomycetidae	伞菌目 Agaricales	Pleurotaceae 侧耳科	香菇属 Lentinula	香菇 (Lentinula edodes)
					侧耳属 Pleurotus	小平菇（Pleurotus cornucopiae）
						平菇 (Pleurotus ostreatus)
						凤尾菇 (Pleurotus pulmonarius)
						金顶侧耳 (Pleurotus citrinipileatus)
						红平菇 (Pleurotus djamer)
						紫孢侧耳 (Pleurotus sapidus)
						杏鲍菇（Pleurotus eryngii）
						鲍鱼菇 (Pleurotus abalonus)
						阿魏蘑 (Pleurotus ferulae)
						盖囊侧耳 (Pleurotus cystidiosus)
						虎奶菇（Pleurotus tuber-regium）

第四章　菌草综合利用
Chapter 4　Comprehensive Application of Juncao

续表

门 Phylum	纲 Class	亚纲 Subclass	目 Order	科 Family	属 Genus	种 Species
担子菌亚门 Basidiomycotina	层菌纲 Hymenomycetes	同隔担子菌亚纲 Holobasidiomycetidae	伞菌目 Agaricales	白蘑科 Tricholomataceae	金针菇属 Flammulina	金针菇（Flammulina velutipes）
						白色金针菇（F. var. velutipes）
					玉蕈属 Hypizygus	玉菇（Hypizygus marmoreus）
						斑玉蕈（Hypsizygus marmoreus）
					口蘑属 Tricholoma	金福菇（Tricholoma lobayensc）
					小奥德蘑属 Oudemansiella	长根菇（Oudemansiella radicata）
					蜜环菌属 Armillariella	蜜环菌（Armillariella mellea）
						假蜜环菌（Armillariella tabescens）
					杯伞属 Clitocybe	大杯蕈（Clitocybe maxima）
				蘑菇科 Agaricaceae	蘑菇属 Agaricus	双孢蘑菇（Agaricus bisporus）
						巴西蘑菇（Agaricus blazei）
						棕色蘑菇（Agaricus brunnescens）
				光柄菇科 Pluteaceae	草菇属 Volvariella	草菇（Volvariella volvacea）
						银丝草菇（Volvariella bombycina）
				球盖菇科 Strophariaceae	鳞伞属 Pholiota	滑菇（Pholiota nameko）
						白滑菇（Pholiota nameko）
					球盖菇属 Stropharia	大球盖菇（Stropharia rugosoannulata）
				鬼伞科 Coprinaceae	鬼伞属 Corprinus	鸡腿菇（Corprinus comatus）
				粪锈伞科 Bolbitiaceae	田头菇属 Agrocybe	柱状田头菇（Agrocybe cylindracea）
						茶树菇（Agrocybe aegerita）

续表

门 Phylum	纲 Class	亚纲 Subclass	目 Order	科 Family	属 Genus	种 Species
担子菌亚门 Basidiomycotina	层菌纲 Hymenomycetes	腹菌亚纲 Gasteromycetidae	鬼笔菌目 Phallales	鬼笔菌科 Phallaceae	竹荪属 Dictyophora	长裙竹荪 (*Dictyophora indusiata*)
						红托竹荪 (*Dictyophora rubrovolvata*)
						棘托竹荪 (*Dictyophora echino-volvata* Zane, Zheng et Hu)
		有隔担子菌亚纲 Phragmobasidiomycetidae	木耳目 Auriculariales	木耳科 Auriculariaceae	木耳属 Auricularia	黑木耳 (*Auricularia auricula*)
						角质木耳 (*Auricularia cornea*)
						毛木耳 (*Auricularia polytricha*)
						盾形木耳 (*Auricularia peltata*)
						皱木耳 (*Auricularia delicata*)
						网状木耳 (*Auricularia auricula*)
						毡盖木耳 (*Auricularia mesenterica*)
			银耳目 Tremellales	银耳科 Tremellaceae	银耳属 Tremella	银耳 (*Tremella fuciformis*)
						金耳 (*Tremella aurantialba*)
						橙耳 (*Tremella cinnabarina*)
		异隔担子菌纲 Heterobasidiomycetes 无隔担子菌亚纲 Holobasidiomycetidae	无褶菌目 Aphyllophorales	灵芝科 Ganodermataceae	灵芝属 Ganoderma	灵芝 (*Ganoderma lucidum*)
						紫芝 (*Ganoderma sinense*)
				多孔菌科 Polyporaceae	革盖菌属 Coriolus	云芝 (*Coriolus versicolor*)
					树花菌属 Grifola	灰树花 (*Grifola frondosa*)
						白树花 (*Grifola albicans*)
					茯苓属 Wolfiporia	茯苓 (*Wolfiporia cocos*)
				猴头菌科 Hericiaceae	枝瑚菌属 Hericium	猴头菌 (*Hericium erinaceus*)

续表

门 Phylum	纲 Class	亚纲 Subclass	目 Order	科 Family	属 Genus	种 Species
担子菌亚门 Basidiomycotina	异隔担子菌纲 Heterobasidiomycetes	无隔担子菌亚纲 Holobasidiomycetidae	无褶菌目 Aphyllophorales	锈革孔菌科 Hymenochaetaceae	针层孔菌属 phellinus	桑黄 (*Phellinus igniarius*)
子囊菌亚门 Ascomycotina	盘菌纲 Discomycete		盘菌目 Pezizales	羊肚菌科 Morchellaceae	羊肚菌属 morchella	羊肚菌 (*Morchella esculenta*)

2. 菌草栽培食药用菌的方法
Method for cultivating edible and medicinal fungi with Juncao grass

根据培养基处理方法不同，菌草栽培食药用菌方式可分为三大类：熟料栽培、发酵料栽培、生料栽培。

The edible and medicinal fungi can be divided into three categories from the treatment method of the culture medium: sterilized material cultivation, fermented material cultivation, and raw material cultivation.

（1）熟料栽培
Cultivation with sterilized substrate

适合栽培的菌类有香菇、毛木耳、黑木耳、平菇、阿魏菇、杏鲍菇、银耳、灵芝、大杯蕈、猴头菌、金针菇、玉菇、滑菇等。有些品种既可用熟料栽培，也可用发酵料和生料栽培，本书介绍的是已规模生产的主要栽培方法，现以栽培香菇工艺流程为例。

Varieties suitable for this cultivation method are: *Lentinula edodes*, *Auricularia polytricha*, *Auricularia auricular*, *Pleurotus ostreatus*, *Pleurotus ferulae*, *Pleurotus eryngii*, *Tremella fuciformis*, *Ganoderma lucidum*, *Clitocybe maxima*, *Hericium erinaceus*, *Flammulina velutipes*, *Hypizygus marmoreus*, *Pholiota nameko*. Some varieties can be cultivated either with sterilized material or raw material. Here introduces the main cultivation methods that have been practiced on a large scale, and now we take the process of cultivating *Lentinula edodes* as an example.

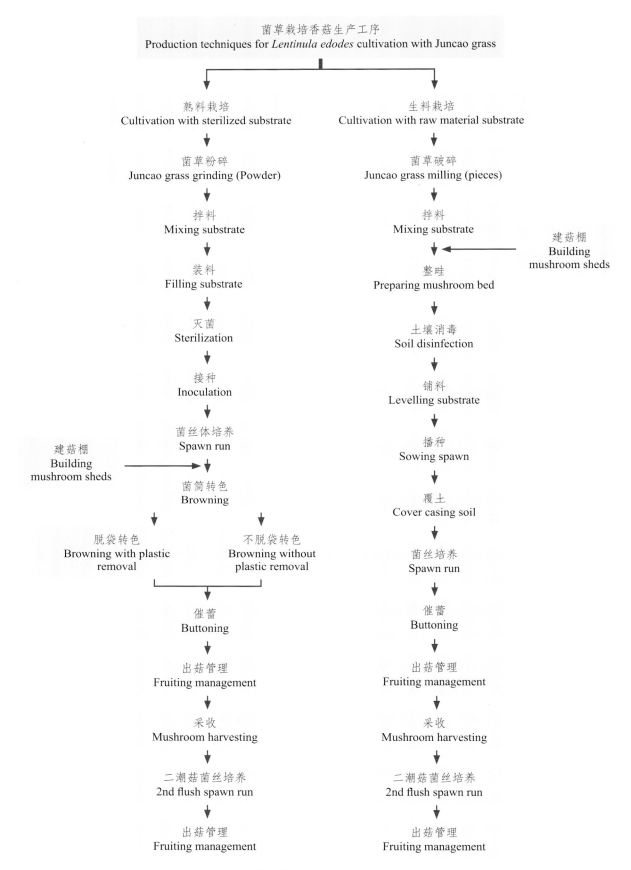

菌草栽培香菇主要工序
The main process of cultivating *Lentinula edodes* by Juncao grass

第四章　菌草综合利用
Chapter 4　Comprehensive Application of Juncao

菌草香菇层架栽培
Cultivation of *Lentinula edodes* with Juncao grass on the shelf

菌草香菇室外地栽
Cultivation of *Lentinula edodes* with Juncao grass on the outdoor ground

菌草墙式栽培毛木耳
Cultivation of *Auricularia polytricha* with Juncao grass for wall-model

菌草袋栽盾形木耳
Cultivation of *Auricularia peltata* with Juncao grass in bags

室外菌草栽培黑木耳
Cultivation of *Auricularia auricula* with Juncao grass outdoors

菌草瓶栽黑木耳
Cultivation of *Auricularia auricula* with Juncao grass in bottles

菌草栽培紫孢平菇
Cultivation of *Pleurotus sapidus* with Juncao grass

菌草栽培黄平菇
Cultivation of *Pleurotus citrinipileatus* with Juncao grass

第四章 菌草综合利用
Chapter 4 Comprehensive Application of Juncao

菌草栽培阿魏菇
Cultivation of *Pleurotus ferulae* with Juncao grass

菌草墙式栽培杏鲍菇
Cultivation of *Pleurotus eryngii* with Juncao grass for wall-model

菌草瓶栽杏鲍菇
Cultivation of *Pleurotus eryngii* with Juncao grass in bottles

菌草袋栽银耳
Cultivation of *Tremella fuciformis* with Juncao grass in bags

菌草墙式栽培灵芝
Cultivation of *Ganoderma lucidum* with Juncao grass for wall-model

菌草栽培鹿角灵芝
Cultivation of *Ganoderma lucidum* of antelope shape with Juncao grass

菌草栽培灵芝
Cultivation of *Ganoderma lucidum* with Juncao grass

菌草栽培猴头菌
Cultivation of *Hericium erinaceus* with Juncao grass

Chapter 4 Comprehensive Application of Juncao

菌草栽培金针菇
Cultivation of *Flammulina velutipes* with Juncao grass

菌草栽培玉菇
Cultivation of *Hypizygus marmoreus* with Juncao grass

菌草瓶栽滑菇
Cultivation of *Pholiota nameko* with Juncao grass in bottles

菌草瓶栽茶薪菇
Cultivation of *Agrocybe aegerita* with Juncao grass in bottles

菌草栽培灰树花
Cultivation of *Grifola frondosa* with Juncao grass

（2）发酵料栽培
Cultivation with composted substrate

适合栽培的种类有双孢蘑菇、棕色蘑菇、巴西蘑菇等。现以菌草栽培双孢蘑菇工艺流程为例。

Varieties suitable for this cultivation method are: *Agaricus bisporus*, *Agaricus brunnescens*, *Agaricus blazei*. Now we take the process of cultivating *Agaricus bisporus* as an example.

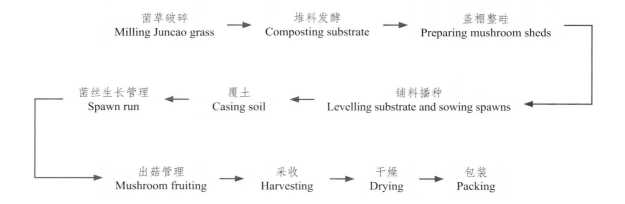

菌草栽培双孢蘑菇主要工序
The main process of cultivating *Agaricus bisporus* by Juncao grass

Chapter 4　Comprehensive Application of Juncao

菌草栽培双孢蘑菇
Cultivation of *Agaricus Bisporus* with Juncao grass

菌草栽培棕色蘑菇
Cultivation of *Agaricus brunnescens* with Juncao grass

菌草栽培巴西蘑菇
Cultivation of *Agaricus blazei* with Juncao grass

（3）生料栽培
Cultivation with raw material substrate

适合栽培的种类有草菇、大球盖菇、竹荪等。现以菌草栽培草菇工艺流程为例。

Varieties suitable for this cultivation method are: *Volvariella volvacea*, *Stropharia rugosoannulata*, and *Dictyophora indusiata*. Now we take the process of cultivating *Volvariella volvacea* as an example.

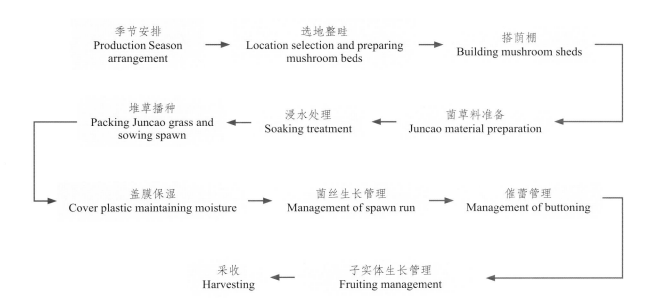

菌草栽培草菇主要工序
The main process of cultivating *Volvariella volvacea* by Juncao grass

菌草栽培草菇
Cultivation of *Volvariella volvacea* with Juncao grass

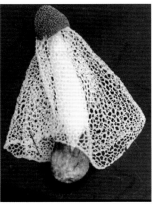

菌草栽培竹荪
Cultivation of *Dictyophora indusiata* with Juncao grass

室外阴棚生料栽培竹荪
Cultivation of *Dictyophora indusiata* with raw material in outdoor shade shed

林地菌草栽培竹荪
Cultivation of *Dictyophora indusiata* with Juncao grass in bamboo forest

菌草室外栽培大球盖菇
Cultivation of *Stropharia rugosoannulata* with Juncao grass outdoors

菌草与菌草菌物饲料

Juncao and Juncao Fungi Forage

巨菌草、绿洲1号等菌草生长快、产量高、营养丰富、适口性好，是牛、羊、马、猪、鹅、鸭、鸡、鱼等家畜、家禽的优质饲料。

Chapter 4　Comprehensive Application of Juncao

With fast growth, high yield, rich nutrition and good palatability, Giant Juncao grass and Lvzhou No.1 are ideal feeds for cattle, sheep, horses, pigs, geese, ducks, chickens, fish and other livestock and poultry.

菌草经微生物发酵，可生产优质高蛋白菌物饲料。
Juncao grass can be fermented by microorganisms to produce high-quality and high-protein fungi feed.

菌草栽培食药用菌的菌糟可直接作饲料或经提取作饲料添加剂。
The spent substrate of edible and medicinal mushroom cultivated by Juncao grass can be used directly as feed or as feed additives after extraction.

菌草灵芝菌糟饲养奶牛，可提高奶牛产奶量，增强奶牛免疫力
Feeding dairy cows with spent substrate of Juncao *Ganoderma lucidum* can increase milk production and improve their immunity.

菌草灵芝的菌糟喂猪，猪血液中的甘油三脂降低 18% 以上，猪肉的胆固醇减低 20% 以上，能有效防治仔猪肠炎痢疾

Feeding the spent substrate of Juncao *Ganoderma lucidum* to pigs reduces the triglycerides in the blood of pigs by more than 18% and the cholesterol of pork by more than 20%, which can effectively prevent piglet enteritis and dysentery

菌草养羊

Raising goats with Juncao grass

第四章　菌草综合利用
Chapter 4　Comprehensive Application of Juncao

菌草养鹿
Raising deer with Juncao grass

菌草养兔
Raising rabbits with Juncao grass

菌草养鹅
Raising geese with Juncao grass

菌草喂鱼
Feeding fish with Juncao grass

菌草养驴
Raising donkeys with Juncao grass

菌草灵芝菌糟提取物（饲料添加剂）
Extraction of spent substrate of Juncao *ganoderma lucidum* (feed additive)

菌草青贮饲料
Juncao silage feed

菌草生物肥料

Juncao Biological Fertilizer

2010年起，开展菌草内生固氮菌及其应用的研究，从巨菌草及绿洲1号的根、茎叶中分离获得具有生防、促生作用的内生固氮菌，且固氮作用强，其中主要菌株为变栖克雷伯氏菌GN02，安全、无毒，可在农业生产中使用。研制了多功能菌草固氮菌肥，可替代25%的化肥，能明显改善小白菜、水稻、巨菌草等作物的生长和品质。

Since 2010, the research on the endogenous nitrogen-fixing bacteria of Juncao grass and its application has been carried out. The endogenous nitrogen-fixing bacteria with biocontrol and growth-promoting effects have been isolated from the roots, stems and leaves of Giant Juncao grass and Lvzhou No.1, which has a strong nitrogen fixation effect. The dominating strain is Klebsiella mutans GN02, safe and non-toxic, applicable in agricultural production. The multifunctional Juncao nitrogen-fixing bacterial fertilizer has been developed, which can replace 25% of chemical fertilizers and significantly improve the growth and quality of crops such as cabbage, rice, and Giant Juncao grass.

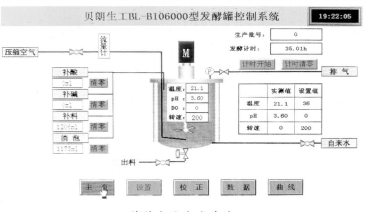

菌草中试生产基地
Juncao Pilot Production Base

菌草生物质能源

Juncao Biomass Energy

1. 菌草发电
Power Generation with Juncao Grass

2008年，在中国浙江兰溪种植巨菌草发电，每公顷巨菌草发电量相当于60吨原煤的发电量
In 2008, Giant Juncao grass was planted in Lanxi, Zhejiang Province, China to generate electricity. The power generation per hectare of Giant Juncao grass was equivalent to 60 tons of raw coal

表 4-2　菌草燃烧热值

Table 4-2　Juncao grass burning heat value

种类 Types	热值（千卡/千克） Calorific value (kcal/kg)	比值 Ratio
原煤 Raw Coal	5000	1
巨菌草 Giant Juncao grass	3580	0.71
菌草平菇菌糟 Spent Substrate of Juncao *Pleurotus ostreatus*	3495	0.7

第四章 菌草综合利用
Chapter 4　Comprehensive Application of Juncao

2. 菌草产沼气
Juncao biogas

每吨巨菌草干物质产沼气量为 548.3 立方米（含 55% 甲烷）
The biogas output per ton of dry matter of Giant Juncao grass is 548.3m^3 (containing 55% methane)

菌草生物质材料

Juncao Biomass Material

2019 年，菌草工业化生产板材
Industrialized production of Fiberboard with Juncao grass in 2019

2021 年，菌草工业化生产中密度纤维板
Industrialized production of medium density Fiberboard with Juncao grass in 2021

2022 年，高质量的菌草刨花板生产性实验成功
In 2022, the production experiment of high-quality Juncao chipboard achieved a success

第五章
助力脱贫攻坚和乡村振兴

Chapter 5
Contribution to Poverty Alleviation and Rural Revitalization

第五章 助力脱贫攻坚和乡村振兴
Chapter 5　Contribution to Poverty Alleviation and Rural Revitalization

1986年菌草技术发明后，首先应用于扶贫领域。自1987年5月起，通过举办技术培训班、建立菌草技术扶贫基地（村、乡、县），召开各类现场会、研讨会等，推动菌草技术扶贫事业的发展。截至2020年，在福建农林大学举办菌草技术扶贫骨干培训班121期，培训来自全国各地的学员6925名。据不完全统计，在项目所在地培训技术人员、扶贫干部和示范生产户9万多人次。

After the successful invention in 1986, Juncao technology was firstly used in the field of poverty alleviation. Since May 1987, Juncao technology was promoted to eradicate poverty by holding technical training courses, establishing demonstration bases (in villages, townships, and counties), and conducting various kinds of on-site meetings and seminars. By 2020, 121 training courses were held at Fujian Agriculture and Forestry University with a total of 6925 trainees from all corners of the country. According to the incomplete statistics, more than 90000 technicians, poverty alleviation officials and demonstration producers have been trained in the different project locations.

利用菌草技术发展菌草业，具有经济效益高、生态效益好、见效快、适应性广、千家万户能参与的优势，先后被列为福建省和国家有关部门重点推广项目。1989年菌草技术被福建省列为科技兴农项目，1990年起被科技部列为"八五""九五""十五"国家级"星火计划"重中之重项目，被中国扶贫基金会列为科技扶贫首选项目，受到广大农村贫困地区、老区、山区干部群众的欢迎。1997年起，菌草技术被福建省列为对口帮扶宁夏、智力援助新疆、对口帮扶三峡库区、科技援藏项目。目前，菌草技术已在我国宁夏、新疆、陕西、内蒙古、贵州、湖北、青海、四川和重庆等31个省（自治区、直辖市）的506个县（市）应用推广，与全国14个集中连片特困区和贫困县共建菌草技术扶贫示范点71个，为促进各地脱贫攻坚和乡村振兴作出积极贡献。

With the advantages of low investment, quick returns, ecological preservation, wide adaptability, and the participation by thousands of households, Juncao technology has been listed as a key promotion by the relevant departments of the state and Fujian Province. Juncao technology was listed as a scientific and technical agriculture project again by Fujian Province in 1989, and as the top priority project of National Spark Program during the Eighth Five-Year plan, Ninth Five-Year plan, and Tenth Five-Year plan periods continuously by the Ministry of Science and Technology in 1990, as the preferred project for poverty alleviation with science and technology by China Foundation for Poverty Alleviation, and widely welcomed by the officials and people in the vast rural poverty-stricken areas, former revolutionary bases, and mountainous places. Since 1997, Juncao technology has been listed by Fujian Province for counterpart assistance project to Ningxia Hui Autonomous Region, intellectual assistance to Xinjiang Uygur Autonomous Region, counterpart assistance to the "Three Gorges" reservoir area, as well as scientific and technical assistance to Tibet. At present, Juncao technology has been applied and promoted to 506 counties in 31 provinces (autonomous regions, municipalities) in China, including Ningxia Hui Autonomous Region, Xinjiang Uygur Autonomous Region, Shaanxi, Inner Mongolia, Guizhou, Hubei, Qinghai, Sichuan and Chongqing. A total of 71 demonstration sites have been established in 14 concentrated contiguous poverty-stricken areas across the country, which have made positive contributions to facilitate poverty alleviation and rural revitalization in various regions.

菌草技术培训

Juncao Technology Training

1988年，由全国农业技术推广总站和福建农学院举办的全国野草栽培食用菌培训班
The National Training Course on Cultivation of Edible Mushrooms with Wild Grass was held by the National Agricultural Technology Extension Station and Fujian Agricultural College in 1988

第五章　助力脱贫攻坚和乡村振兴
Chapter 5　Contribution to Poverty Alleviation and Rural Revitalization

1993年3月，福建农学院第36期菌草技术骨干培训班
In March 1993, the 36th Juncao Technology Backbones Training Course of Fujian Agricultural College

1996年8月，在江西宁冈县开展菌草技术扶贫培训班
In August 1996, the Juncao Technology Poverty Alleviation Training Course in Ninggang County, Jiangxi Province

1998年4月，福建农业大学与宁夏科委在彭阳县举办六盘山区培训班
In April 1998, Fujian Agricultural University and the Science and Technology Commission of Ningxia Hui Autonomous Region co-organized the Liupan Mountain Training Course in Pengyang County

2003年12月，西部地区菌草技术骨干培训班
In December 2003, the Juncao Technology Backbones Training Course for the Western Region

Chapter 5 Contribution to Poverty Alleviation and Rural Revitalization

2006年11月,在宁夏召开菌草产业扶贫现场观摩培训会
On-site observation training seminar for poverty alleviation with Juncao industry was held in Ningxia Hui Autonomous Region in November 2006

2011年11月,新疆昌吉回族自治州菌草技术骨干培训班
In November 2011, the Juncao Technology Backbones Training Course in Xinjiang Changji Hui Autonomous Prefecture

2013年10月,菌草技术骨干培训班
The Juncao Technology Backbones Training Course in October 2013

2016年1月，菌草产业扶贫培训班
The Juncao Industry Poverty Alleviation Training Course in January 2016

2016年11~12月，贵州黔西南州菌草技术骨干培训班
The Juncao Technology Backbones Training Course in Qianxinan Prefecture, Guizhou Province
in November-December 2016

菌草技术示范基地

Juncao Technology Demonstration Bases

1. 福建将乐
In Jiangle of Fujian Province

将乐菌草灵芝林间菇棚（2003~2021 年）
Juncao *Ganoderma lucidum* shed in Jiangle (2003-2021)

2. 福建顺昌
In Shunchang of Fujian Province

福建顺昌畲族村发展菌草新型产业助力乡村振兴
The She Nationality Village in Shunchang county of Fujian Province develops Juncao industry to beef-up rural revitalization

第五章 助力脱贫攻坚和乡村振兴
Chapter 5 Contribution to Poverty Alleviation and Rural Revitalization

3. 福建泰宁
In Taining of Fujian Province

泰宁种植菌草治理水土流失，同时用菌草饲养家禽家畜
Juncao grass was planted in Taining County to control soil erosion and raise poultry and livestock

4. 福建福安
In Fu' an of Fujian Province

福安发展菌—草—畜新型菌草产业，助力乡村振兴
Develop new Juncao industry of mushrooms-grass-livestock to facilitate rural revitalization

5. 宁夏彭阳
In Pengyang of Ningxia Hui Autonomous Region

菌草技术被列为闽宁协作对口帮扶项目，种植菌草发展菌草菇助力脱贫攻坚，成效显著

Juncao technology was listed as a counterpart assistance project of Fujian-Ningxia Cooperation with remarkable effect for poverty elimination

6. 新疆
In Xinjiang Uygur Autonomous Region

福建农林大学在福建省、昌吉回族自治州有关部门支持下，菌草技术援疆成效显著，菌草业成为农牧民增收的特色产业（2001~2021 年）

With the support of the relevant departments of Fujian Province and Changji Hui Autonomous Prefecture of Xinjiang Uygur Autonomous Region, Fujian Agriculture and Forestry University has scored remarkable achievements in its technical assistance to Xinjiang. Juncao has become a specialty industry for farmers and herdsmen to increase their income (in 2001-2021)

7. 西藏
In Tibetan Autonomous Region

菌草技术列为科技部与福建省部省会商项目，为高原藏区的生态治理和藏民脱贫攻坚开辟了新途径（2011~2021年）

Juncao technology is listed as a consultative cooperation project of the Ministry of Science and Technology and Fujian Province, which has opened up a new way for the ecological management of Tibetan plateau areas and the poverty alleviation for Tibetans (in 2011-2021)

8. 内蒙古
In Inner Mongolia Autonomous Region

第五章 助力脱贫攻坚和乡村振兴
Chapter 5 Contribution to Poverty Alleviation and Rural Revitalization

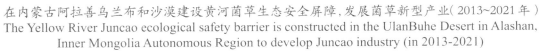

在内蒙古阿拉善乌兰布和沙漠建设黄河菌草生态安全屏障，发展菌草新型产业（2013~2021年）
The Yellow River Juncao ecological safety barrier is constructed in the UlanBuhe Desert in Alashan, Inner Mongolia Autonomous Region to develop Juncao industry (in 2013-2021)

9. 贵州
In Guizhou Province

Chapter 5　Contribution to Poverty Alleviation and Rural Revitalization

在贵州种植巨菌草治理石漠化，获得成功（1996~2021年）
Giant Juncao grass planted to control rocky desertification in Guizhou Province achieved great success (in 1996-2021)

10. 青海
In Qinghai Province

第五章 助力脱贫攻坚和乡村振兴
Chapter 5　Contribution to Poverty Alleviation and Rural Revitalization

在青海贵德黄河上游高寒地区进行菌草生态安全屏障建设研究示范的成功，为高寒地区生态安全和新型产业发展开辟了新途径（2014~2021年）

The successful studies and demonstration for the construction of Juncao ecological safety barrier in the alpine region of the upper reaches of the Yellow River in Guide county, Qinghai Province, has opened up a new way for ecological security and development of new industries in the alpine region (in 2014 -2021)

11. 四川
In Sichuan Province

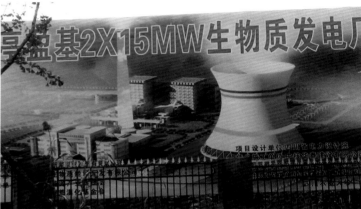

1998年，四川省扶贫办建立菌草开发研究中心发展菌草，助力脱贫攻坚
In 1998, the Sichuan Provincial Poverty Alleviation Office established the Juncao Development and Research Center to promote poverty alleviation

全国菌草扶贫会议

The National Symposiums on Poverty Alleviation with Juncao Technology

1995年3月27日，在福建农林大学召开福建省菌草业发展战略学术研讨会
On March 27, 1995, Fujian Provincial Juncao Industry Development Strategy Symposium was held at Fujian Agriculture and Forestry University

1997年5月20~23日，在北京召开全国第一届菌草技术扶贫研讨会
The First National Poverty Alleviation Symposium on Juncao Technology was held in Beijing on May 20-23, 1997

1998年11月2~4日，在福建农林大学召开全国第二届菌草技术扶贫研讨会
The Second National Poverty Alleviation Symposium on Juncao Technology was held at Fujian Agriculture and Forestry University on November 2-4, 1998

2000年6月22~24日，在宁夏银川召开全国第三届菌草技术扶贫研讨会
The Third National Poverty Alleviation Symposium on Juncao Technology was held in Yinchuan, Ningxia Hui Autonomous Region on June 22-24, 2000

2002年10月29日，在新疆召开全国第四届菌草技术暨天山菌草产业发展研讨会
The Fourth National Symposium of Juncao Technology&Tianshan Juncao Industry Development was held in Xinjiang Uygur Autonomous Region on October 29, 2002

2004年10月23~24日，在福州召开全国第五届菌草技术扶贫研讨会
The Fifth National Poverty Alleviation Symposium on Juncao Technology was held in Fuzhou on October 23-24, 2004

2005年9月13~16日，在宁夏银川召开第六届菌草技术扶贫研讨会
The Sixth Poverty Alleviation Symposium on Juncao Technology was held in Yinchuan, Ningxia Hui Autonomous Region on September 13-16, 2005

第五章　助力脱贫攻坚和乡村振兴
Chapter 5　Contribution to Poverty Alleviation and Rural Revitalization

2008年1月12日，在福建农林大学召开菌草技术扶贫和援外工作十年总结表彰大会
On Januray 12, 2008, the Ten-year Summary Conference on Juncao Technology Poverty Alleviation and Foreign Aid was held at Fujian Agriculture and Forestry University

2017年8月24日，在延安召开全国菌草产业发展与精准扶贫现场会
The National Conference on Juncao Industry Development and Targeted Poverty Alleviation was held in Yan'an on August 24, 2017

2018年10月13日，在贵州召开中国菌草产业发展与精准扶贫研讨会
On October 13, 2018, the Symposium on Juncao Industry Derelopment and Targeted Poverty Alleviation was held in Guizhou Province

2020年6月17日，在贵州贵阳召开菌草生态修复与产业扶贫座谈会
The Symposium on Juncao Ecological Restoration and Industrial Poverty Alleviation was held in Guiyang, Guizhou Province on June 17, 2020

菌草创新产业园

Juncao Innovation Industrial Park

1. 宁夏石嘴山闽宁协作菌草科技创新产业园
The Fujian-Ningxia Collaborative Juncao Technology Innovation Industrial Park in Shizuishan City, Ningxia Hui Autonomous Region

在宁夏石嘴山盐碱地、河滩地种植菌草，发展"动物—植物—菌物"三物循环的菌草新型生态产业，提高村民和退役军人自我发展能力，助力石嘴山乡村振兴、黄河宁夏段菌草生态屏障建设。

Juncao grass was planted in the saline-alkali land and river beaches of Shizuishan Municipality, Ningxia Hui Autonomous Region, and a new type of Juncao ecological industry with the circular production of "animal-plant-mushroom" was developed, which improved the self-development capabilities of villagers and veterans, helped the revitalization of Shizuishan villages and the construction of Juncao ecological barriers in the Ningxia section of the Yellow River basin.

2021年4月17日，黄河宁夏菌草科技创新产业园项目启动仪式在石嘴山市平罗县举行
On April 17, 2021, the launching ceremony of the Yellow River Ningxia Juncao Technology Innovation Industrial Park project was held in Pingluo County, Shizuishan Municipality

菌草工厂化育苗
Workshop-scale Breeding of Juncao Grass Seedlings

第五章　助力脱贫攻坚和乡村振兴
Chapter 5　Contribution to Poverty Alleviation and Rural Revitalization

菌草机械化种植
Mechanized Planting of Juncao Grass

2. 内蒙古磴口菌草科技创新产业园
Juncao Technology Innovation Industrial Park in Dengkou County, Inner Mongolia Autonomous Region

2021年6月1日，在内蒙古巴彦淖尔市磴口县召开菌草科技创新产业园启动仪式。利用磴口黄河滩地和乌兰布和沙漠的沙地种植菌草，发展畜牧业、菌业、加工业，实现一草多用、综合利用、循环利用。

On June 1, 2021, the launching ceremony of Juncao Technology Innovation Industrial Park was held in Dengkou County, Bayannur Municipality, Inner Mongolia Autonomous Region. Bottom-land of the

Yellow River in Dengkou County and the sandy land of the Ulan Buh Desert are used to grow Juncao grass, develop animal husbandry, mushroom industry, and processing industries, to realize multiple functions, comprehensive utilization and recycling of Juncao grass.

On June 1, 2021, the launching ceremony of the Juncao Science and Technology Innovation Industrial Park was held in Dengkou County, Inner Mongolia Autonomous Region

3. 河南武陟黄河生态菌草科技创新产业园
Yellow River Ecological Juncao Technology Innovation Industrial Park in Wuzhi County, Henan Province

武陟县位于黄河中下游分界处，是黄河地上悬河的起点。2021年6月18日，在河南省焦作市武陟县召开"菌草科技创新产业园"启动大会。通过菌草科技创新产业园的建设，在武陟黄河滩地种植万亩菌草，建黄河菌草生态安全屏障，发展菌草畜业、菌草菌业、加工业，实现"植物—菌物—动物"三物对资源综合循环高效利用、生态绿色可持续发展，助力乡村振兴。

Wuzhi County is located at the boundary of the middle and lower reaches of the Yellow River and is the starting point of the suspended river above the Yellow River. On June 18, 2021, the Juncao Technology Innovation Industrial Park kick-off meeting was convened in Wuzhi County, Jiaozuo municipality, Henan Province. Through the establishment of Juncao Technology Innovation Industrial Park, 10,000 acres of Juncao grass was planted on the Wuzhi Yellow River basin, for building up a Juncao ecological safety barrier for the Yellow River, developing Juncao livestock industry, Juncao mushroom industry, and downstream processing industry, thus realizing the plants-mushroom-animals recycling with more efficient utilization of resources, promoting sustainable development and rural revitalization.

2021年6月18日，河南武陟"菌草科技创新产业园"启动，走产业生态化、生态产业化的绿色发展之路
On June 18, 2021, the Juncao Science and Technology Innovation Industrial Park in Wuzhi County, Henan Province was launched, taking the green development road of industrial ecology and ecological industry

第六章 服务国际减贫和可持续发展

Chapter 6
Promotion of International Poverty Reduction and Sustainable Develepment

国际交流合作

International Exchange and Cooperation

为了加强国际技术交流与合作，发展菌草业，造福全人类，自1996年11月以来，在国内外举办了19届菌草业发展国际研讨会。

In order to strengthen the international technical exchanges and cooperation to develop Juncao industry and benefit all mankind, 19 international symposiums on the development of Juncao industry have been held in China and abroad since November 1996.

1. 国际交流
International Exchange

1992年4月，菌草技术项目参加第20届日内瓦国际发明竞赛，国际评委认为该技术解决了"菌林矛盾"和"菌粮矛盾"难题，为人类提供优质菇类食品，为畜牧业发展提供优质饲料，开辟了一条最经济最合理的途径，获日内瓦州政府奖和金奖

In April 1992, the Juncao technology project participated and won awards at the 20th Geneva International Exhibition of Inventions. The international judges agreed that the Juncao technology solved the contradictions between "Mushroom cultivation-Forest reservation " and the "mushroom cultivation-Food production ", which has opened up the most economical and reasonable way to provide high-quality mushroom food for humans and feed for developing animal husbandry. Juncao technology scored both the Geneva State Government Award and the Gold Award

1994年5月，菌草栽培食用菌综合技术参加法国巴黎第85届国际发明竞赛，国际评委认为该技术是国土整治和开发普遍适用的技术，获法国内政部国土整治规划部奖

In May 1994, the integrated technology of Juncao cultivation for edible fungi participated in the 85th French International Invention Competition in Paris. The international judges considered this a universally applicable technology for territorial governance and development, and won the award of the French Interior Ministry and the Ministry of Territorial governance and Planning

1996年11月25~29日，在中国福州召开首届菌草业发展国际研讨会，达成"发展菌草业，造福全人类"的共识

The first International Symposium on Juncao Industry Development was held in Fuzhou, China from November 25-29, 1996, and the consensus of "Developing Juncao Industry for the Benefit of All Mankind" was reached

2000年12月6日，在巴西的巴西利亚召开菌草食用菌国际会议
On December 6, 2000, the International Conference of Juncao Edible Fungi was held in Brasilia, Brazil

第六章 服务国际减贫和可持续发展
Chapter 6　Promotion of International Poverty Reduction and Sustainable Development

2005年9月13~16日，在中国宁夏银川召开第三届国际菌草业发展暨第六届中国菌草技术扶贫研讨会
The Third International Symposium on Juncao Industry Development was held on September 13-16, 2005 in Yingchuan, Ningxia Hui Autonomous Region, China

2006年11月17日，在南非彼得马里茨堡召开第四届菌草业发展国际研讨会
The Fourth International Symposium on Juncao Industry Development was held in Pietermaritzburg, South Africa on November 17, 2006

2014年6月，在福州召开第十二届国际菌草产业发展研讨会
The 12th International Juncao Industry Development Symposium was conducted in Fuzhou in June 2014

2015年，17国驻华使节在世界菌草技术发源地福建农林大学种植菌草纪念
In 2015, the ambassadors and diplomatic envoys of 17 countries to China planted Juncao grass for commemoration in Fujian Agriculture and Foresty University (FAFU), the birthplace of Juncao technology

2017年6月，第十五届国际菌草产业发展研讨会暨院士专家论坛在福州召开
The 15th International Symposium and Academicians Experts Forum on Juncao Industry Development was held in Fuzhou in June 2017

第六章 服务国际减贫和可持续发展
Chapter 6 Promotion of International Poverty Reduction and Sustainable Development

2019年6月，第十七届国际菌草产业发展研讨会暨院士专家论坛在福州召开
The 17ᵗʰ International Symposium and Academicians Experts Forum on Juncao Industry Development was held in Fuzhou in June 2019

2. 国际合作
International Cooperation

(1) 日本 Japan

1992年，"提高菌草培养料营养的栽培食用菌方法"的发明专利在日本的使用权转让给日本加贺市中国贸易开发中心株式会社。

In 1992, the invention patent of Method of Cultivating Edible Mushrooms with Improved Nutrition of Juncao Substrate was transferred to the China Trade Development Center Co., Ltd. in Kaga City, Japan.

1992年9月25日，福建农学院与日本"中国贸易开发中心株式会社"举行签字仪式
On September 25, 1992, Fujian Agricultural College and the China Trade Development Center Co., Ltd. in Japan held a signing ceremony for the transfer

1993年4~9月，林占熺（右二）、吴兆辉（右一）、林占华（左一）菌草技术专家赴日本石川县加贺市共建"石川菌草茸研究所"
From April to September 1993, Chinese Juncao Technology experts Lin Zhanxi (second from right), Wu Zhaohui (first from right), and Lin Zhanhua (first from left) jointly established the "Ishikawa Juncao Mushroom Research Institute" in Kaga City, Ishikawa Prefecture, Japan

菌草瓶栽滑菇
Cultivation of *Pholiota nameko* with Juncao substrate in bottles

菌草瓶栽小平菇（姬平菇）
Pleurotus ostreatus cultivated with Juncao substrate in bottles

菌草瓶栽猴头菌
Hericium erinaceus cultivated with Juncao substrate in bottles

菌草瓶栽毛木耳
Auricularia polytricha cultivated with Juncao substrate in bottles

2003年5月至2004年8月，福建农林大学与日本宫城县仙台市梦实耕望会社合作建立菌草栽培香菇示范基地（右一为福建农林大学专家林兴生）

From May 2003 to August 2004, Fujian Agriculture and Forestry University cooperated with the Espuire company in Sendai City, Miyagi County, Japan to build the Juncao Cultivation Shiitake Mushroom Demonstration Base (Mr. Lin Xingsheng, expert from Fujian Agriculture and Forestry University, first from right)

（2）巴西 Brazil

1995年10月，巴西科学院教授艾瑞德·乌本参加在福州举办的首届菌草技术国际培训班，引进技术、菌种和资料。并在圣保罗建立了菌草菇示范基地，与福建农林大学合作举办菌草技术培训班、菌草菇国际会议，推动了南美洲的菌草业发展。

In October 1995, Professor Arailde Urben of Embrapa (Brazilian Academy of Sciences) participated in the first International Training Course on Juncao Technology held in Fuzhou. She introduced Juncao technology, strains, substrate materials to Brazil, and established a Juncao Mushroom Demonstration Base in Sao Paulo in cooperation with Fujian Agriculture and Forestry University. The Juncao technical training course and the Juncao Mushroom International Conference promoted the development of Juncao industry in South America.

1995年10月，艾瑞德·乌本教授参加首届菌草技术国际培训班，并种植菌草留念

In October 1995, Professor Arailde Urben participated in the first International Course of Juncao Technology and planted Juncao grass for commemoration

第六章 服务国际减贫和可持续发展
Chapter 6 Promotion of International Poverty Reduction and Sustainable Develpement

2002年12月起，在巴西巴西利亚、圣保罗等地召开6次菌草食用菌国际会议
Since December 2002, six international conferences on mushroom have been convened in Brasilia, Sao Paulo, and other cities in Brazil

（3）埃及 Egypt

2000年5月13日，林占熺应埃及农业部邀请赴埃及考察，研究组建沙漠菌草产业研究与示范基地
As invited by the Egyptian Ministry of Agriculture, Professor Lin Zhanxi visited Egypt on May 13, 2000 for studies on establishing the Research and Demonstration Base for Desert Juncao Industry in Egypt

2012年11月，林占熺和林冬梅在埃及沙漠研究所作菌草治沙与新型菌草业的研究报告
In November 2012, Lin Zhanxi and Lin Dongmei made a research report on Sand Control with Juncao and New Juncao Industry in the Egyptian Desert Research Institute

（4）朝鲜 DPRK

1995年，朝鲜派专家参加首届国际菌草技术培训班，引进菌草技术，发展菌草菇生产。2016年引进巨菌草发展菌草畜业。

In 1995, DPRK sent experts to participate in the first international training course of Juncao technology, and introduced the technology to promote the production of Juncao mushrooms. In 2016, the Giant Juncao grass was introduced to develop Juncao livestock industry.

第六章　服务国际减贫和可持续发展
Chapter 6　Promotion of International Poverty Reduction and Sustainable Development

2016年9月30日至10月4日，林占熺赴朝鲜介绍中国菌草技术扶贫，并指导巨菌草种植
From September 30 to October 4, 2016, Professor Lin Zhanxi made a visit to DPRK to introduce Chinese Juncao technology for poverty alleviation and gave technical guidance on the planting of Giant Juncao grass

（5）缅甸 Myanmar

2015年，在缅甸果敢地区试种了约 6.7 公顷菌草，2016 年种了 100 公顷菌草。利用菌草养牛、羊、兔、猪、鸡、鸭、鹅，帮助当地农民脱贫，弃种罂粟，取得了很好的效果，受到当地农民和执政者的欢迎。

In 2015, about 6.7 hectares grass of Juncao were planted experimentally in the Kokang area of Myanmar. In 2016, Juncao grass was expanded to 100 hectares to help local farmers raising cattle, sheep, rabbits, pigs, chickens, ducks, and geese to get rid of poverty and abandon poppy cultivation. It has achieved excellent results and much welcomed by the local farmers and authorities.

缅甸果敢地区发展菌草业替代罂粟种植
Juncao industry is developed in the Kokang area of Myanmar to replace poppy cultivation

人才培养

Talent Fostering

1. 菌草技术国际培训班
Juncao Technology International Training Course

1995 年，首届菌草技术国际培训班
The First International Training Course on Juncao Technology in 1995

1997 年，菌草技术国际培训班
International Training Course on Juncao Technology in 1997

1999年，菌草技术国际培训班
International Training Course on Juncao Technology in 1999

2000年，菌草技术国际培训班
International Training Course on Juncao Technology in 2000

第六章 服务国际减贫和可持续发展
Chapter 6　Promotion of International Poverty Reduction and Sustainable Develepment

2001年，菌草技术国际培训班
International Training Course on Juncao Technology in 2001

2005年，菌草技术国际培训班
International Training Course on Juncao Technology in 2005

2006年，发展中国家菌草技术培训班
Training Course on Juncao Technology for Developing Countries in 2006

2007年，菌草技术国际培训班
International Training Course on Juncao Technology in 2007

第六章 服务国际减贫和可持续发展
Chapter 6 Promotion of International Poverty Reduction and Sustainable Developement

2008年，菌草技术国际培训班
International Training Course on Juncao Technology in 2008

2009年，菌草技术国际培训班
International Training Course on Juncao Technology in 2009

2010年，发展中国家菌草技术培训班
International Training Course on Juncao Technology for Developing Countries in 2010

2011年，发展中国家农业产业发展官员研修班
Agricultural Industry Development Seminar for Officials from Developing Countries in 2011

2012年，发展中国家菌草产业发展官员研修班
Juncao Industry Development Seminar for Officials from Developing Countries in 2012

第六章 服务国际减贫和可持续发展
Chapter 6　Promotion of International Poverty Reduction and Sustainable Develepment

2012年，发展中国家菌草产业发展官员研修班
Juncao Industry Development Seminar for Officials from Developing Countries in 2012

2013年，发展中国家菌草技术培训班结业典礼
Closing Ceremony of Juncao Technology Training Course for Developing Countries in 2013

2014年，菌草技术国际培训班
International Training Course on Juncao Technology in 2014

2014年，非洲法语国家菌草技术培训班
Juncao Technical Training Course for French-speaking African Countries in 2014

2015年，菌草技术国际培训班
International Training Course on Juncao Technology in 2015

第六章 服务国际减贫和可持续发展
Chapter 6 Promotion of International Poverty Reduction and Sustainable Developement

2016年，发展中国家菌草技术培训班
International Training Course on Juncao Technology for Developing Countries in 2016

2017年，国际菌草技术培训班
International Training Course on Juncao Technology in 2017

2018 年，非洲法语国家菌草产业发展官员研修班
Juncao Industry Development Seminar for Officials from French-speaking African Countries in 2018

2019 年 11 月，菌草技术国际培训班
International Training Course on Juncao Technology in November 2019

第六章 服务国际减贫和可持续发展
Chapter 6 Promotion of International Poverty Reduction and Sustainable Develepment

2011年6月28日，中国驻埃及大使宋爱国和驻厄立特里亚大使李连生到福建农林大学看望参加发展中国家菌草技术培训班的学员
On June 28, 2011, the Chinese Ambassador to Egypt Song Aiguo and the Chinese Ambassador to Eritrea Li Liansheng visited Fujian Agriculture and Forestry University (FAFU) to call on the trainees from Egypt and Ethiopia who participated in the Training Course on Juncao Technology for Developing Countries

2011年12月8日，卢旺达驻华大使恩加兰贝到福建农林大学看望参加发展中国家菌草技术培训班的学员
On December 8, 2011, the Rwandan Ambassador to China H. E. Francois Xavier Ngarambé visited FAFU to call on the trainees from Rwanda who participated in the Training Course on Juncao Technology for Developing Countries

2019年7月11日,国家国际发展合作署署长王晓涛莅临福建农林大学调研菌草技术援外工作,参观中非学员的学习成果

On July 11, 2019, H.E. Wang Xiaotao, Director General of China International Development Cooperation Agency, visited FAFU for studies on the assistance work with Juncao technology and the gains in of the trainees from Central African Republic

2019年中非菌草技术海外培训班开班仪式

Opening Ceremony of Juncao Technology Training Course was held in Central African Republic in 2019

第六章　服务国际减贫和可持续发展
Chapter 6　Promotion of International Poverty Reduction and Sustainable Development

2014年中国援斐济菌草技术合作项目第二期菌草技术培训班
The 2nd Training Course held by China-aid Fiji Juncao Technical Cooperation Project in 2014

在卢旺达举办菌草技术海外培训班
Training Course on Juncao Technology was held in Rwanda

中国援巴布亚新几内亚菌草、旱稻技术援助项目培训班
Training Course held by China-Aid Papua New Guinea Juncao and Upland Rice Technical Assistance Project

2016年6月29日，斐济总统乔治·孔罗特给学员颁发结业证书
On June 29, 2016, George Konrote, President of Fiji, issued certificates of completion to the trainees

第六章　服务国际减贫和可持续发展
Chapter 6　Promotion of International Poverty Reduction and Sustainable Develepment

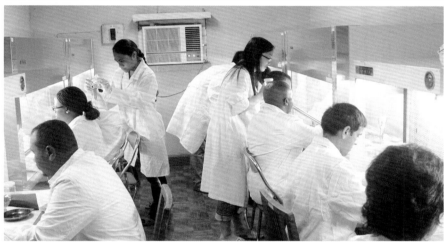

学员实践课
Student practice classes

学员留言
Student Message

十几个国家的专家学者一起种植菌草，象征着友谊，预示着造福全人类的菌草业的崛起。

1995 年 10 月

Planting Juncao grass by the experts coming from fourteen countries is the symbol of friendship. It predicts the rise of Juncao industry which will benefit humanity.

In October, 1995

我们相信，菌草技术将是食品安全和自然环境保护的世界发展趋势的最佳选择。——来自 14 个国家的 23 名第二届国际菌草培训班（1997 年 10 月 8 日 ~11 月 3 日）学员。

We believe that Juncao Technology will be the best option for the developing world towards food security and the conservation on natural environment. —— By 23 participants from 14 countries of 2nd International Juncao Training Course organized at Fuzhou(October 8- November 3,1997).

我们，第三届菌草培训班（1997 年 11 月 12 日 ~12 月 1 日）的学员也许不能够在时代的沙滩中留下自己的脚印，但是我们将在为祖国和人类服务中努力发扬菌草精神，传授菌草技术，实现菌草梦想。

高效的营养转换与自然资源的有效保护将保证菌草技术在全球的绿地上菌业的发展。

We, the participants of the 3rd Juncao workshop (November 12 - December 1, 1997) might not be leaving our footprints in the sands of time but we will strive to promote the philosophy, teach the skill and implement the vision of Juncao in the service of our nations and humanity.

Efficient nutrients conversion with natural resources conservation will ensure that Juncao endeavors will mushroom in all green places on earth.

2. 硕、博士生培养
Cultivation of Master and Ph.D Students

菌草中心自 2000 年开始在国内招收菌草专业硕、博士生，2009 年开始招收留学生，至今共培养了中国、阿富汗、莱索托、卢旺达、南非、尼日利亚、巴勒斯坦、加纳、埃及、坦桑尼亚、肯尼亚、马来西亚等 12 个国家的 129 名硕、博士生。

The Juncao Center has started to receive master and doctoral students in China since 2000, and international students since 2009. So far, 129 master and doctoral students have been cultivated, they are from China, Afghanistan, Lesotho, Rwanda, South Africa, Nigeria, Palestine, Ghana, Egypt, Tanzania, Kenya, and Malaysia.

菌草中心教师与国内外学生庆祝元旦
Teachers and students of the Juncao Center celebrate the New Year together

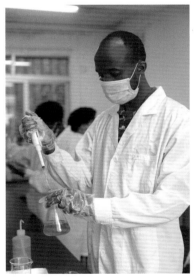 

留学生进行实验课题研究
Juncao international students conduct experimental research

林占熺带领中外硕博士生实践——菌草治理崩岗
Lin Zhanxi leads the practice of Juncao students——collapsed hills control with Juncao grass

林占熺与2017届毕业生及在读硕博士生合影留念
Lin Zhanxi took a group photo with 2017 Juncao graduates and some currently enrolled students

中国援建菌草技术示范中心

Juncao Technology Demonstration Centers Established with Aid from China

1. 巴布亚新几内亚
Papua New Guinea

1997年5月，福建省与巴布亚新几内亚东高地省签署了菌草技术重演示范协议。2000年5月，时任福建省省长习近平与巴布亚新几内亚东高地省省长拉法纳玛签署《福建省援助东高地省发展菌草、旱稻生产技术项目协议书》和《友好省合作协议》。2018年11月，习近平总书记见证了中国援助巴布亚新几内亚菌草和旱稻项目的签署。通过建立菌草技术示范基地，把技术性较强、小农户不易参与的如菌草加工、菌种和菌袋生产，集中在示范基地进行，由中国专家带领经过培训的技术员和村民来完成。把菌草种植、收割，搭建菇棚，菌袋出菇管理等技术简化后，由小农户参与，成功地解决了偏僻山区的贫困村民运用现代菌草技术脱贫的难题。从当地气候、土壤等实际条件出发，发明了旱稻宿根法栽培技术，在菌草种植方面创造了巨菌草一年产鲜草853吨/公顷的纪录，在旱稻栽培方面创造了播种一次连续收割13季和干谷最高产量达8.5吨/公顷的纪录，近年发展菌草养猪、养羊等。菌草与旱稻两项技术从东高地省辐射至巴布亚新几内亚全国，受益农户超过6000户，参加菌草菇和旱稻生产的农户增加了收入，解决了稻米自给问题。对增加就业、减贫和粮食安全起积极作用。

In May 1997, Fujian Province of China and Eastern Highlands Province of Papua New Guinea (hereinafter referred to as EHP-PNG) signed an agreement on the demonstration of Juncao technology in EHP-PNG. In May 2000, H.E. Xi Jinping, then Governor of Fujian province and H.E. Rafalama, Governor of EHP, signed a *Cooperation Agreement on Juncao and Upland Rice Technology Projects with Economic Aid by Fujian Province to EHP-PNG* and *Sisterly-relations Accord* between the twin provinces. In November 2018, General Secretary Xi Jinping witnessed the signing of Juncao and Upland Rice Project with China aid to PNG. Through the establishment of the Juncao technology demonstration base, the highly technical and difficult procedures operations, such as the processing of Juncao, the production of strains and fungi bags, will be concentrated in the demonstration base and completed by technicians and staff led by Chinese experts. After simplification of Juncao grass planting, harvesting, mushroom shed building, mushroom bag fruiting management, the smallholder farmers and poverty-stricken villagers in remote mountainous areas could be able to apply modern Juncao technology to get rid of poverty. Based on the actual local climate and soil conditions, the upland rice after-growth technique was invented. The Giant Juncao grass harvest created a record-breaking yield of 853 tons per hectare of fresh grass annually. The upland rice cultivation has also set a record of 13 continuous harvests by planting at one time and growing for many seasons with the highest yield of dry grain reaching 8.5 tons per hectare. Recent years has seen the development of pig and sheep raising with Juncao grass as forage. The Juncao and upland rice technologies extended from Eastern Highlands province to many places of PNG, benefiting more than 6000 farmer households. The farmers who participated in the production of Juncao mushrooms and upland rice drastically increased their income and achieved self-sufficiency in rice, which plays a positive role in increased employment, better food security and reduced poverty.

在巴布亚新几内亚东高地省鲁法区建立的首个菌草技术示范基地
The first Juncao Technology Demonstration Base established in Lufa District, EHP - PNG

第六章 服务国际减贫和可持续发展
Chapter 6 Promotion of International Poverty Reduction and Sustainable Develepment

1998年1月14日，巴布亚新几内亚在东高地省鲁法区召开全国菌草技术示范成功现场会。全国各地5000多人参加会议，巴布亚新几内亚总督、副总理和巴布亚新几内亚国家8位部长出席会议

A national Juncao technology demonstration mass rally was held in Lufa District, Eastern Highlands Province, Papua New Guinea on January 14,1998. More than 5000 people from all over the country, the Governor of Papua New Guinea, the Deputy Prime Minister and eight ministers of PNG attended the festival-styled rally

1998年1月14日，巴布亚新几内亚总督西拉斯·阿图培尔视察菌草技术示范生产基地

Sir Silas Atopare (center), Governor General of PNG, made an inspecting tour to Juncao Technology Demonstration Base accompanied by Lin Zhanxi (right) and Lin Yuexin (left) on January 14, 1998

2001年5月31日，巴布亚新几内亚首相莫劳塔访问福建农业大学并发表演讲

H.E. Mekere Morauta, Prime Minister of Papua New Guinea, visited FAFU and delivered a speech at the welcome ceremony on May 31, 2001

2001年5月31日，巴布亚新几内亚首相莫劳塔访问福建农林大学，与中国驻巴布亚新几内亚大使赵振宇、福建省副省长潘心城一同种植象征友谊的菌草

H. E. Mekere Morauta, Prime Minister of PNG, and H.E. Zhao Zhenyu, Chinese Ambassador to PNG, and H.E. Pan Xincheng, then Deputy Governor of Fujian province, were planting Juncao grass, as a symbol of friendship on May 31, 2001

第六章 服务国际减贫和可持续发展
Chapter 6 Promotion of International Poverty Reduction and Sustainable Develpement

1998年8月30日，中国驻巴布亚新几内亚大使张鹏翔把在巴布亚新几内亚首次栽培的菌草平菇赠送给国会议员、劳工部部长马西亚斯·卡拉尼

H.E. Zhang Pengxiang, the Chinese Ambassador to PNG, presented the Juncao oyster mushroom firstly cultivated in Papua New Guinea to H.E. Mathias Kalani, MP and Minister of Labor of PNG on August 30, 1998

1999年10月，中国科学院院士谢联辉（右一）等到鲁法菌草技术示范基地考察指导

Prof. Xie Lianhui (first from right), an Academician of the Chinese Academy of Sciences, visited the Lufa Juncao Technology Demonstration Base for inspection and guidance in October 1999

2000年7月31日，中国驻巴布亚新几内亚大使赵振宇（前排左三）、经商处参赞刘连科（前排左一）到鲁法菌草技术示范基地视察指导

H.E. Zhao Zhenyu, Chinese Ambassador to PNG (third from left in from row), and H. E. Liu Lianke (first from left in from row), Economic & Commercial Counselor made an inspection visit to Lufa Juncao Technology Demonstration Base on July 31, 2000

2003年3月6日，巴布亚新几内亚总督西拉斯·阿图培尔向第五期菌草技术培训班的学员颁发结业证书
H. E. Silas Atopare, Governor General of PNG, delivered the completion certificate to a participant of the fifth Juncao technical training course on March 6, 2003

2003年3月6日，房志民参赞代表我国政府向巴布亚新几内亚东高地省赠送农机具
H. E. Fang Zhimin, Economic &Commercial Counselor, presented agricultural hand tools as gifts to EHP-PNG on behalf of the Chinese government on March 6, 2003

2005年5月，福建农林大学王豫生率代表团访问巴布亚新几内亚，参加援巴新菌草旱稻技术培训班结业典礼
A delegation led by Mr. Wang Yusheng, then head of FAFU, made a visit to PNG and attended the training course graduation ceremony of China-Aid Juncao and Upland Rice Technology Project in May 2005

第六章 服务国际减贫和可持续发展
Chapter 6 Promotion of International Poverty Reduction and Sustainable Develepment

2014年6月18日，巴布亚新几内亚东高地省省长朱莉索叶访问福建，在第12届国际菌草产业发展研讨会上介绍巴布亚新几内亚菌草、旱稻项目

H.E. Julie Sawyer, then Governor of EHP - PNG, made a presentation on the PNG Juncao and Upland Rice Project during her visit to Fujian to attend the 12th International Juncao Industry Development Symposium on June 18, 2014

2017年8月17日，中国驻巴布亚新几内亚经商参赞刘林林（左一）代表我国政府赠送东高地省菌草、旱稻生产机具和设备

H. E. Liu Linlin (left), Economic and Commercial Counselor of the Chinese Embassy in PNG, presented Juncao and upland rice production machines and facilitiesto EHP on behalf of the Chinese government on August 17, 2017

2018年4月15日，东高地省省长彼得·努姆（右二）、中国驻巴布亚新几内亚大使薛冰（左三）出席庆祝旱稻丰收大会

H.E. Peter Numu (second from right), Governor of EHP and H.E. Xue Bing (third from left), Chinese Ambassador to PNG, attended the celebration functions for bumper harvest of upland rice on April 25, 2018

2019年11月1日，福建省副省长郭宁宁（左四）访问巴布亚新几内亚，考察由中国援助的菌草技术与旱稻技术项目

H.E. Guo Ningning (fourth from left), Vice Governor of Fujian Province made an inspection visit to the Juncao and Upland rice technology project aided by China, during her visit to PNG on November 1, 2019

2019年12月27日，巴布亚新几内亚总理詹姆斯·马拉佩（前排左七）夫妇视察中国援助巴新菌草技术与旱稻技术项目示范基地

H. E. James Marape (seventh from left in front row), Prime Minister of PNG and his wife made an inspection tour to the China-Aid Juncao Technology and Upland rice Project on December 27, 2019

第六章 服务国际减贫和可持续发展
Chapter 6　Promotion of International Poverty Reduction and Sustainable Develepment

1998 年，菌草生料栽培平菇
Cultivation of *Pleurotus ostreatus* with Juncao raw material in 1998

1999 年，林间菌草栽培紫孢平菇
Cultivation of *Pleurotus ostreatus* with Juncao grass in the forest in 1999

1999 年，鲁法小学学生参加种植菌草实践
The students from Lufa Primary School participated in Juncao grass planting

2017年11月23日，巴布亚新几内亚东高地省和西高地省农业厅在高路卡菌草技术示范基地组织测产，巨菌草年产鲜草达每公顷853吨

The Agricultural Departments of EHP and Western Highlands Province of PNG organized a yield estimation on November 23, 2017 at the Goroka Juncao Technology Demonstration Base, and the annual output of fresh Giant Juncao grass reached 853 tons per hectare

一批菇农通过菌草生产住上了新房，改善了生活

The local farmers have drastically improved their livelihood through Juncao mushroom production, many of them now live in their newly constructed houses

2. 南非
South Africa

2003年2月，南非祖鲁王国古德维尔·兹韦利蒂尼国王首次访问福建农林大学。2004年2月，福建农林大学与夸祖鲁－纳塔尔省合作进行菌草技术重演示范实验，获得成功后于7月由福建农林大学与夸祖鲁－纳塔尔省农业厅签署在南非建"菌草旱稻技术示范基地"合作协议，2005年1月项目开始实施。

In February 2003, H. E. Majesty King Goodwell Zwelithini of the Kingdom of KwaZulu in South Africa visited Fujian Agriculture and Forestry University (FAFU) for the first time. In February 2004, FAFU in cooperation with KwaZulu-Natal Province conducted a demonstration experiment on Juncao technology. With the success, FAFU and the KwaZulu-Natal Provincial Department of Agriculture concluded a contract in July to establish a "Juncao & Upland Rice Technology Demonstration Base" in South Africa, and the project began to be implemented in January 2005.

2006年12月13日，福建省省长黄小晶与夸祖鲁－纳塔尔省省长恩德贝勒签署两省结成姐妹省及菌草旱稻合作协议。

On December 13, 2006, H.E. Huang Xiaojing, then Governor of Fujian Province, and H.E. Ndebele, Governor of KwaZulu-Natal Province, signed a Sisterly Relations and Juncao and Upland Rice Cooperation Agreement between the two provinces.

该项目实施旨在通过发展菌草产业助力南非贫困农户脱贫，特别是帮助"穷人中的穷人"脱贫。项目实施创建了"菌草示范中心＋旗舰点（卫星基地）＋示范点（合作社）"的模式，建立了1个研究培训中心，7个旗舰点，超过40个示范点，并创新总结出"四化四结合"经验，即技术"本土化、简便化、标准化和系统化"，项目实施运行与当地需求、自然条件、村民、政府相结合，解决了发展中国家农业技术项目进村入户难和持续发展难两大难题。当地受益人数超过万人，为当地粮食安全、提高贫困农户收入、创造就业作出贡献。

The implementation of the project aims to help poor farmers in South Africa get rid of poverty through the development of Juncao industry, especially to help "the poorest among the poor". The implementation of the project created a model of "Juncao Demonstration Center + Flagship Site (Satellite Base) + Demonstration Site (Cooperative)", established one research and training center, seven flagship sites, more than 40 demonstration sites. It summarized "the successful" experience, that is, the technology "localization, simplification, standardization and systematization", and integation the project implementation and operation with local needs, natural conditions, villagers, and governments. It effectively solved the two major problems of household participation and sustainable development in promoting agricultural technology in rural areas of developing countries. The number of local beneficiaries exceeded 10000, contributed to local food security, increasing the income of poor farmers, and creating new employment opportunities.

2003年2月18日，南非祖鲁王国古德维尔·兹韦利蒂尼国王访问福建农林大学
H. M. King Goodwell Zwelithini of Zulu Kingdom of South Africa made a visit to FAFU on February 18, 2003

2003年9月19日，南非夸祖鲁-纳塔尔省交通部部长 Sibusiso Ndebele（左四），南非国会议员黄士豪（左三）在彼得马里茨堡会见福建农林大学代表团
The Premier of KwaZulu-Natal Province, South Africa, H.E. Sibusiso Ndebele (fourth from left) and MP Dr. Charlie Huang (third from left) met with FAFU delegation in Pietermaritzburg on September 19, 2003

2005年11月30日，南非祖鲁王国古德维尔·兹韦利蒂尼国王和黄士豪议员再次访问福建农林大学，并种植菌草留念
H. M. King Goodwell Zwelithini of the Zulu Kingdom of South Africa and MP Dr. Charlie Huang, plant Giant Juncao grass as a memento on November 11, 2005 on their second visit to FAFU

第六章 服务国际减贫和可持续发展
Chapter 6 Promotion of International Poverty Reduction and Sustainable Developement

2005 年 11 月，南非祖鲁王国古德维尔·兹韦利蒂尼国王访问福建农林大学，并看望参加菌草技术国际培训班的南非学员

H. M. king Goodwill Zwelithini, of Zulu Kindom of South Africa called on the South African participants at FAFU training seminar in November 2005

南非菌草技术研究示范中心位于南非夸祖鲁－纳塔尔省省府彼得马里茨堡市西德拉镇，占地 2 公顷，具有科学研究、示范、技术培训、良种繁育等功能

South Africa Juncao Technology Research and Demonstration Center is located in Cedera Town, Pietermaritzburg, the capital of KwaZulu-Natal Province. It covers an area of two hectares and has the functions of scientific research, demonstration, technical training, and breeding

原料车间
Inoculation Workshop

生产车间
Production Workshop

接种车间
Raw Material Workshop

办公楼
Office Building

培训教室与示范车间
Training Rooms and Demonstration Workshop

第六章 服务国际减贫和可持续发展
Chapter 6　Promotion of International Poverty Reduction and Sustainable Develepment

根据菌草产业化扶贫和可持续发展的目标，开创了"中心＋旗舰点（卫星基地）＋示范点"的实施模式，形成专业化分工，实行专业合作社管理，有效降低农户的技术和市场风险。

In accordance with the goals of Juncao industrialization for poverty alleviation and sustainable development, the implementation model of "Center + Flagship Site (satellite base) + Demonstration Site" was created to form a specialized division of labor, implemented professional cooperative management, and effectively reduced farmers' technical and market risks.

夸丁地菌草旗舰点
Juncao flagship site in KwaDingdi

夸丁地农场是首个菌草旗舰点，2010年10月完工并投入使用，拥有106个合作社成员。2008年7月29日，中国驻南非钟建华大使（右三）向夸丁地旗舰点赠送菌草生产机器设备

KwaDingdi Farm is the first Juncao flagship site with 106 cooperative members, which was completed and put into operation in October 2010. H.E. Zhong Jianhua (third from right), the Chinese Ambassador to South Africa presents Juncao production equipment to the flagship site on July 29, 2008

Dukuduku 旗舰点
Dukuduku flagship site

Nxamalala 旗舰点
Nxamalala flagship site

Enyokeni 旗舰点
Enyokeni flagship site

第六章　服务国际减贫和可持续发展
Chapter 6　Promotion of International Poverty Reduction and Sustainable Develepment

2005年10月26日，福建省副省长王美香与南非夸祖鲁－纳塔尔省省长恩德贝勒在彼得马里茨堡市签署两省建立友好省份关系意向书

Deputy Governor of Fujian Province H.E. Wang Meixiang and Governor of KwaZulu-Natal Province H.E. Sibusiso Ndebele signed a memorandum of friendship between the two provinces in Pietermaritzburg on October 26, 2005

2006年11月21日，中国驻南非大使刘贵今出席在夸丁地旗舰点举行的菌草项目启动大会

The Chinese Ambassador to South Africa H.E. Liu Guijin attended the Juncao project launching ceremony held at the flagship site of KwaDingdi on November 21, 2006

2006年12月31日，福建省省长黄小晶、南非夸祖鲁－纳塔尔省省长恩德贝勒在福州签署建立友好省关系协议书

Governor Huang Xiaojing of Fujian and Premier Sibusiso Ndebele of KwaZulu-Natal Province signed the Agreement on Friendship Province Relationship in Fuzhou on December 13, 2006

2013年3月16日，福建农林大学校长兰思仁率领代表团考察南非夸祖鲁-纳塔尔省西德拉菌草技术中心
Prof. Lan Siren, President of FAFU, led a delegation for an inspection visit to the Cedera Juncao Technology Center in KwaZulu-Natal Province, South Africa on March 6, 2013

2013年3月25日，南非祖鲁王国国王古德维尔·兹韦利蒂尼在德班接见由福建农林大学党委书记叶辉玲（左三）率领的福建农林大学代表团
H. M. King Goodwill Zwelithini received the delegation from FAFU led by Ye Huiling (third from left), the Party Committee of FAFU in Durban on March 25, 2013

（1）在南非举办各类型菌草技术培训班
Various Type of Juncao Technology Training Courses Held for South Africa

2005年，南非夸祖鲁-纳塔尔省农业厅研究人员在西德拉接受培训
Researchers from the Department of Agriculture of KwaZulu-Natal Province, received training in Cedera in 2005

Chapter 6　Promotion of International Poverty Reduction and Sustainable Develepment

南非夸祖鲁-纳塔尔省农业厅研究人员尼尔（左三）和夸丁地菌草菇合作社社长保罗（右三）在福建农林大学学习，并获得结业证书

Mr. Neil (third from left), the Department of Agriculture KZN researcher, and Paul (third from right), the Director of KwaDingdi Juncao Mushroom Growers' Cooperative received the Completion Certificates in FAFU training course

（2）开展适应性试验示范
Adaptive Test Demonstration Conducted

2004年，在南非夸祖鲁-纳塔尔省彼得马里茨堡市种植巨菌草，生长期1年，高3.24米，每公顷产鲜草450吨

The Giant Juncao grass planted in Pietermaritzburg, KwaZulu-Natal Province, South Africa, with a growth period of 1 year, a height of 3.24 meters, and fresh grass production of 450 tons per hectare in 2004

2005 年 3 月 8 日，首次在南非马卡梯尼滨海沙地种植巨菌草，94 天高 2.2 米
On March 8, 2005, the Giant Juncao grass was planted in the coastal sands of Makatini, South Africa for the first time, with a height of 2.2 meters in 94 days

机械化种植巨菌草，6 道工序一次完成，3 人 1 小时种 1 公顷
Machine planting of Giant Juncao grass: finishing 6 procedures at one time, 3 people planting 1 hectare in 1 hour

南非菌草养羊
Goat farming with Juncao grass as forage in South Africa

（3）菌草扶贫模式
The Model of Poverty Alleviation with Juncao

创建10平方米菇房年产鲜菇1.2吨的菌草扶贫模式，破解南非"穷人中的穷人"——单亲妈妈脱贫难题
A 10-square-meter mushroom shed was created with an annual output of 1.2 tons of fresh mushrooms. This model of poverty alleviation with Juncao is to help single mothers who are "the poorest among the poor" in South Africa up and out of poverty

2006年4月，南非夸祖鲁-纳塔尔省山区贫困单亲母亲喜收菌草平菇
The single mothers in mountainous area of KwaZulu-Natal Province, happily celebrate their bumper harvest of the new crop in April 2006

艾滋病患者之家9平方米的菇棚，为5个艾滋病患者全年提供菇类食品
The 9-square-meter mushroom shed of the home of AIDS patients provides mushroom food for 5 patients annually

3. 卢旺达
Rwanda

2008年5月2日，中国驻卢旺达大使孙树忠与卢旺达国际合作署部长签署《关于中国援助卢旺达农业技术示范中心合作议定书》。该中心位于尼罗河上游卢旺达南方省胡也区国家农科院鲁伯纳实验站，占地面积22公顷。中心以应用菌草技术发展菌草业为特色，包括水土保持、稻作种植和蚕桑养殖等，采用"中心+旗舰点+农户"的模式，在卢旺达首都基加利和各省各设立1个旗舰点。

On May 2, 2008, H.E. Sun Shuzhong, the Chinese Ambassador to Rwanda, and the Minister of Rwanda International Cooperation Agency signed the Protocol of Cooperation on China's Assistance to the Rwanda Agricultural Technology Demonstration Center. The center is located at the Rubona Experimental Station of the National Academy of Agricultural Sciences, Huye District, Southern Province in Rwanda, on the upper reaches of the Nile River, covering an area of 22 hectares. The center is characterized by the application of Juncao technology to develop Juncao industry, including soil and water conservation, rice planting and sericulture, etc., adopting the model of "center + flagship site + farmer households", and set up one flagship site in Kigali, the capital of Rwanda, and to each of the provinces.

"中国援助卢旺达农业技术示范中心"于2009年4月2日开工，2011年4月1日竣工。应卢旺达的要求，为了更快发挥中心作用，采用边建设边示范培训和运行的方法，使示范中心提早2年开始进入运行。

The construction of the China Aid Rwanda Agricultural Technology Demonstration Center was started on April 2, 2009 and completed on April 1, 2011. Upon the request of Rwanda side, in order to play the role of the center faster, demonstration and training were conducted alongside the construction, so that the demonstration center actually started operation two years earlier.

根据卢旺达在水土流失、气候、土壤、作物种类等方面的特点，中国专家研究出投入少、见效快、小农户能参与的水土流失治理新技术，建立起菌草与玉米、大豆、花生、番薯、果树、茶叶、桑树等作物间作模式，开辟了把水土保持和卢旺达传统农业生产相结合，实现良性循环持续发展新途径。

Based on Rwanda's characteristics in soil erosion, climate, soil, crop types, etc., Chinese experts have developed new technologies for soil erosion control with low investment, quick results, and easy for small farmers' participation, and established the intercropping of Juncao grass with corn, soybean, peanut, sweet potato, fruit trees, tea, mulberry and other crops, opened up a new way of combining soil and water conservation with traditional agricultural production in Rwanda and achieved a virtuous cycle of sustainable development.

第六章 服务国际减贫和可持续发展
Chapter 6　Promotion of International Poverty Reduction and Sustainable Develepment

参与项目的农户收入大幅度增加，其中仅菌草菇一项已经有超过 3500 个农户收入增加一倍以上，破解了就业难、脱贫难问题，取得显著成效，为卢旺达就业减贫和可持续发展发挥了积极作用。

The income of farmers participating in the project increased significantly. The income of more than 3,500 farmers from mushroom production was more than doubled, which has solved the problem of unemployment and achieved remarkable results. It has contributed to poverty reduction and sustainable development in Rwanda with a positive effect.

中国援卢旺达农业技术示范中心
China Aid Rwanda Agricultural Technology Demonstration Center

2012 年 4 月 24 日，卢旺达总理皮埃尔·达米安·哈布姆兰伊（左一）为中国援卢旺达农业技术示范中心正式启动剪彩。他说："我赞赏中国给予卢旺达的援助和友谊，非常感谢它将使我国人民未来的生活更加美好！"
The Prime Minister of Rwanda, H.E. Pierre Damian Habumranyi (first from left), cut the ribbon for the official launching of the China Aid Rwanda Agricultural Technology Demonstration Center on April 24, 2012. He said: "I appreciate China's assistance and friendship to Rwanda, and I am very grateful that it will make our people's lives better in the future!"

2010年10月21日，卢旺达外交与合作部部长路易丝·穆希基瓦博视察中国援卢旺达农业技术示范中心
后题词："卢旺达外交部和政府诚挚感谢中国对卢旺达人民创造更加美好生活的贡献。"
The Minister of Foreign Affairs and Cooperation of Rwanda, H.E. Louise Mujikiwabo, visited to the China Aid Rwanda Agricultural Technology Demonstration Center on October 21, 2010. She wrote an inscription saying "The Ministry of Foreign Affairs and the government of Rwanda sincerely thank China for its contribution to the creation of a better life for the Rwandan people."

2012年10月18日，卢旺达农牧资源部部长卡莉巴塔说："我们很感谢农业技术示范中心所做的工作，我认为这会成为技术转移的桥梁，从中国引入卢旺达没有的技术，他们推广技术的方式也很新颖，而且这种技术转移也可以真正让百姓学得会，我们很感激。"
"We really appreciate what is doing. So I think that it's going to bridge the gap between technology transfer, the technology from China that we don't have here. The means and the ways that reaching to farmer is so new, so the technology transfer, the knowledge to reach farmer is something we really appreciate." (CCTV News, interview to Hon. Dr. Agnes Kalibate, Minister of MINAGRI, Rwanda, on October 18, 2012)

第六章 服务国际减贫和可持续发展
Chapter 6 Promotion of International Poverty Reduction and Sustainable Development

2017年3月31日，卢旺达总统卡加梅（前排右三）在总统府接见福建农林大学国家菌草中心副主任林冬梅（前排左一），对中国援卢旺达农业示范中心项目的合作表示支持

H.E. President Paul Kagame of Rwanda (third from right, front row) received Dr. Lin Dongmei (first from left, front row), the Deputy Director of FAFU National Juncao Research Center in President's Office. During the conversation, he expressed strong support to C-RATDC project on March 31, 2017

（1）水土保持
Soil and Water Conservancy

在尼罗河上游种植菌草用于水土流失治理效果好、见效快，与传统作物玉米比，种植巨菌草的土壤流失率减少97%以上。

Juncao grass in the upper reaches of the Nile River has a good effect on soil erosion control and takes effect quickly. Compared with the traditional crop corn, the soil loss rate after planting Giant Juncao grass is reduced by more than 97%.

卢旺达总理皮埃尔·达米安·哈布姆兰伊视察中心菌草水土保持试验点

H.E. Dr. Pierre Damien HABUMUREMYI, Prime Minister of the Republic of Rwanda made an inspection visit to the soil and water conservation experiment site in the Center

水土流失治理试验
Experiment on soil and water loss control

巨菌草与玉米、大豆、花生、番薯、果树、茶叶、桑树等传统作物间作，水土保持和传统农业生产相结合，简便实用，实现菌草、作物、畜业三物良性循环，可持续发展
Intercropping of Giant Juncao grass with traditional crops such as corn, soybean, peanut, sweet potato, fruit trees, tea, mulberry, and combining water and soil conservation with traditional agricultural production, is simple and practical, and achieves a virtuous cycle of Juncao grass, crops, and animal husbandry for sustainable development

（2）菌草菇生产
Juncao Mushroom Production

示范推广菌草技术，引进草与菌的新品种和栽培技术，使菌草技术本土化和简便化，让农户易掌握。

Demonstrate and extend Juncao technology originated in China, introduce new varieties of Juncao grasses and mushrooms, and its cultivation techniques, and to adapt and simplify the technology for farmers to learn easily.

在卢旺达种植巨菌草。当地运用菌草技术栽培平菇，每平方米菇床一年能产鲜菇 120 千克
In Rwanda, mushroom bed per square metre can produce 120kg fresh mushroom in a year with Juncao technology

指导孤儿院师生参加菌草菇生产
Instruct teachers and students of the orphanage to produce Juncao mushrooms

指导敬老院种植菌草菇
Instruct people to grow Juncao mushrooms in nursing homes

指导妇女协会（大屠杀遗孀）种植菌草菇
Instruct women's Association (Widows of the Massacre) to grow Juncao mushrooms

第六章 服务国际减贫和可持续发展
Chapter 6　Promotion of International Poverty Reduction and Sustainable Develepment

大学毕业生栽培菌草食用菌创业
University graduates start a business by cultivating Juncao edible mushroom

2011 年 5 月 21 日，在卢旺达举办的首期培训班结业
Completion of the first training course on May 21, 2011

2015年1月15~19日为NYAMAGABE地区妇女菌草菇合作社举办第34期菌草技术培训班
The 34th Training Course on Juncao Technology for the NYAMAGABE Women Mushroom Cooperative was held on January 15-19, 2015

2016年1月26~29日，中国专家到基伍湖岛青少年管教所为1900多名青少年示范、指导种菇
On January 26-29, 2016, the Chinese experts conducted training course on mushroom cultivation at IWAWA island prison, Kivu lake for more than 1900 young inmates, accompanied by local agricultural experts

4. 莱索托
Lesotho

2006年10月9日，中国政府与莱索托政府签订换文协定，把菌草技术列为中国援助莱索托项目，中国商务部把该项任务下达给福建农林大学组织实施。

On October 9, 2006, the Chinese government and the Lesotho government signed a Letter of Exchange, listing Juncao technology as a Chinese aid project to Lesotho, and the Ministry of Commerce assigned the task to Fujian Agriculture and Forestry University (FAFU) to undertake the implementation.

项目的目标是：推动莱索托农户发展新兴菌草业，解决莱索托发展畜牧业的饲料短缺问题，为莱索托水土流失治理提供新措施、开辟新途径。高效利用水、土地资源，实现植物、菌物和动物三物对资源综合循环利用，实现可持续发展。

The project objective is to promote the development of the emerging Juncao industry among the farmers in Lesotho, solve the problem of feed shortage in animal husbandry, and provide new ways for control of soil erosion in Lesotho, so as to make efficient utilization of water and land resources, realize the comprehensive recycling of resources by plants, fungi and animals to enhance sustainable development.

2007年9月，中国菌草技术专家组抵达莱索托，开始执行援莱索托菌草技术项目，到2021年5月，完成了四期的援莱索托各项任务，将中莱菌草技术示范基地建成菌草技术试验、示范、培训、推广基地，为当地增加就业、消除贫困和食品安全提供新途径。

In September 2007, the Chinese Juncao technical expert team arrived in Lesotho and began to implement the Juncao technology project. By May 2021, altogether four phases of the aid project were completed The China-Lesotho Juncao Technology Demonstration Base was established with functions of technical experiment, demonstration, training, and promotion, and provided new ways for increased local employment, poverty eradication and food safety.

根据莱索托的需要和生产条件，项目的实施采用"示范基地+旗舰点+合作社+农户"的方式，破解了技术进村到户难和持续发展难的问题，为增加就业、减贫发挥了积极作用。

Based on Lesotho's needs and the practical production conditions, the implementation of the project adopts the model of "demonstration base + flagship site + cooperative + farmer households" which effectively solved the two major problems in household participation and sustainable development in promoting agricultural technology in villages of Lesotho, and played an important role in creating job opportunities and poverty reduction with positive effects.

2007年10月18日，莱索托首相帕卡利塔·莫西西利（中）访问福建
Lesotho Prime Minister H.E. Pakalitha Mosisili (middle) made a visit to Fujian province on October 18, 2007

2005年，莱索托农业部部长玛帕莱莎·莫托霍女士（前排左二）访问福建农林大学时提出引进菌草技术
Ms. Mapalesa Mothokho (second from left, front row), Minister of Agriculture of Lesotho in 2005 visited FAFU and proposed to introduce Juncao Technology to Lesotho

2008年6月5日，莱索托国王、王后、副首相及中国驻莱大使仇伯华到莱索托全国农展会参观菌草技术项目，并充分肯定菌草技术对莱索托减贫的作用
Their Majesties the King and the Queen of the Kingdom of Lesotho, and the Deputy Prime Minister, accompanied by the Chinese Ambassador H.E. Qiu Bohua for a visit to Juncao project exhibition stall and tasted mushroom cuisine on June 5, 2008

第六章　服务国际减贫和可持续发展
Chapter 6　Promotion of International Poverty Reduction and Sustainable Develepment

2009年4月8日，高德毅大使举办"菌草菇宴"，向莱索托各部委介绍中国援助莱索托菌草技术项目

H.E. Gao Deyi the Chinese Ambassador, held a "mushroom feast" to promote the awareness of Juncao technology among the high-ranking officials of the ministries in Lesotho on April 8, 2009

2018年2月9日，在中国驻莱索托大使馆举办中国援莱索托菌草技术合作项目推介会。孙祥华大使，宋常清政务参赞，马国良经商参赞，莱索托小企业大臣波里，农业部代大臣马夸伊，莱索托外交部、发展计划部等相关政府部门官员，以及当地各界代表出席

On February 9, 2018, a promotion meeting of China aid Juncao technology cooperation project was in session at the Chinese Embassy in Lesotho. The Ambassador H.E. Sun Xianghua, the Political Counselor Mr. Song Changqing, the Economic and Commercial Counselor Mr. Ma Guoliang, the Minister of Small Business H.E. Pori, the Acting Minister of Agriculture H.E. Ma Kwayi, the officials from the Ministry of Foreign Affairs, Ministry of Development and Planning, and other relevant government departments and local representatives from all walks of life attended the event

（1）培养菌草学科留学生
Students from Lesotho majoring in Juncao technology nurtured

2010年9月至2013年7月，马洛塔·撒科迪在福建农林大学学习菌草技术
Mr. Malota Sekete studied Juncao Technology in FAFU from September 2010 to July 2013

2012年9月至2015年7月，科内尔·拉麦撒在福建农林大学学习菌草技术
Mr. Kornel Ramaisa studied Juncao Technology in FAFU from September 2012 to July 2015

2015年9月至2018年7月，马西塔·玛梅罗·莫科迪在福建农林大学学习菌草技术
Ms. Masita Mamello Mokete studied Juncao Technology in FAFU from September 2015 to July 2018

（2）举办各种类型培训班
Various Types of Training Courses Held

2007年，在中莱菌草技术示范基地为莱索托农业部技术人员和项目官员举办菌草技术培训班
In China-Lesotho Juncao Technology Demonstration Base, a training course on Juncao technology was held for the technicians and project officials of the Ministry of Agriculture in Lesotho in 2007

2010年5月20日，为当地的非政府组织（NGO）举办菌草技术培训班
Holding a Juncao technology training courses for NGOs on May 20, 2010

2014年3月19日，在加查斯内克区哈－瑟卡科村举办菌草技术培训班
The training Course on Juncao Technology was held in Ha-Sekake Village of Qacha's Nek District in Lesotho on March 19, 2014

2014年4月2日,在马塞卢区马博莱卡纳村举办菌草技术培训班
The Juncao Technology Training Course was held in Mabolekana Village of Maseru on April 2, 2014

2014年8月28日,在贝瑞亚区哈-利措拉村举办菌草技术培训班
The Juncao Technology Training Course was held in Ha-Letsoela village of Berea District on August 28, 2014

2015年5月11日,在莱索托马塞卢职业培训中心为残疾人农户举办菌草技术培训班
The Juncao Technology Training Course was held for disabled farmers at Maseru Vocational Training Center in Lesotho on May 11, 2015

2011年3月25日，为莱索托国立罗马大学举办菌草技术培训班
The Juncao Technology Training Course was held for Rome National University of Lesotho on March 25, 2011

2015年1月29日，为莱索托马西安康高中学生举办菌草技术培训班
The Juncao Technology Training Course was held for MASIANOKENG high school students in Lesotho on Jonuary 29, 2015

为莱索托农学院学生举办菌草技术培训班
The Juncao Technology Training Course was held for Lesotho Agriculture College students

（3）组建菌草生产者协会
Formation of Juncao Producers Association

莱索托第一个菌草生产协会
The first Farmers' Association for Juncao production in Lesotho

莱索托第二个菌草生产协会
The second Farmers' Association for Juncao production in Lesotho

（4）建旗舰点
Establishment of Flagship Sites

在加查斯内克区、马塞卢区、贝瑞亚区和撒巴－泽卡地区共建设了 8 个菌草技术旗舰点。项目基地为各旗舰点提供原材料，并进行技术示范指导。

A total of 8 Juncao technology flagship sites were set up in Gachasneck District, Maseru District, Berea District and Saba-Zeka District respectively. The Juncao Demonstration Base provided raw materials for each flagship site, and conduct demonstration and technical guidance.

第六章 服务国际减贫和可持续发展
Chapter 6 Promotion of International Poverty Reduction and Sustainable Develepment

2014 年 4 月 2 日，马塞卢区马博莱卡纳村菌草技术旗舰点
Mushroom shed in Juncao Technology Flagship Site in Mabolekana Village of Maseru on April 2, 2014

2014 年 10 月 9 日，贝瑞亚区哈－拉齐岛村菌草技术旗舰点
Juncao Technology Flagship Site in Ha-Ratsiu Village in Berea District on October 9, 2014

2014 年 11 月 27 日，塔巴－齐卡区菌草技术旗舰点
Juncao Technology Flagship Site in Thaba-Tsika District on November 27, 2014

2014年12月12日，马西安康建菌草技术旗舰点来种草养畜
Juncao Technology Flagship Site in Masianokeng for grass planting and livestock breeding on December 12, 2014

2020~2021年，发生新冠肺炎疫情，中国菌草技术专家组依托示范基地，坚持技术推广，持续发挥技术优势，在马塞卢新建菌草菇菌袋生产旗舰点，扩大生产规模和持续发展
During the 2020-2021 COVID-19 pandemic period, the Chinese Juncao technical expert team at the demonstration base, insisted on technology promotion, continued to provide expertise services, and even built a new flagship site for Juncao mushroom bag production in Maseru, for expanding production scale and promoting sustainable development

第六章 服务国际减贫和可持续发展
Chapter 6 Promotion of International Poverty Reduction and Sustainable Development

2014年3月18日，加查斯内克菌草养羊旗舰点
Demonstration of sheep feeding with Juncao grass in the Flagship Site in Ha-Sekake Village of Qacha's Nek District in Lesotho on March 18, 2014

5. 斐济
Fiji

中国援助斐济菌草技术示范中心项目，是由中斐两国领导人共同确定、推动的技术援助项目。2013年11月7日，中国和斐济两国签署关于援斐济菌草技术合作换文协议，该项目由福建农林大学负责实施，通过合作研究、技术培训、示范生产，建立各种类型的示范点、示范户，促进斐济菌草新型生态产业的发展。

The China-Fiji Juncao Technology Demonstration Center Project is a technical assistance project jointly determined and promoted by the Heads of State of China and Fiji. On November 7, 2013, China and Fiji signed a Letter of Exchange on Juncao Technology Cooperation as a China aid project to Fiji, which is implemented by Fujian Agriculture and Forestry University (hereinafter referred to as FAFU) who carried out through cooperation in studies, technical training, and demonstration production, setting up various types of demonstration sites, demonstration households to facilitate the development of Juncao production in Fiji as a new ecological industry.

中国援斐济菌草技术示范中心设有菌草良种繁育示范种植区8公顷，杧果园套栽菌草食药用菌2公顷，建年产鲜菇300吨的菌草食用菌生产线，引进菌草饲料收割、粉碎、打包、青贮等生产设备。

The China-Fiji Juncao Technology Demonstration Center has a land of 8 hectares as Juncao grass seedlings breeding demonstration area, 2 hectares for inter-planting edible and medicinal mushrooms in the mango garden, established a production line of Juncao edible mushroom with an annual output of 300 tons of fresh mushrooms, and introduced Juncao forage harvesting, grinding, packing, silage and other production equipment.

经中斐双方多年努力，示范中心项目取得了显著成效，通过技术培训与示范推广，已向全国提供500公顷巨菌草种苗，缓解西部干旱季节缺乏饲料的难题；结束了当地没有食药用菌生产的历史，参加菌草菇生产的农户超过600户，参加种植巨菌草发展菌草畜业生产农户超过1000户。菌草新型生态产业已成为斐济的特色产业，为斐济等海岛国家增加就业、减贫和应对气候变化与可持续发展开辟一条社会效益大、生态效益好、经济效益高的新途径。

With years of efforts by both the Chinese and Fijian sides, the Demonstration Center project has made remarkable achievements. Through technical training, demonstration and promotion, 500 hectares of Giant Juncao grass seedlings have been provided to the country to alleviate the problem of lack of feed in dry season the west part of the country; ended the history of lack of edible and medicinal mushrooms cultivation in Fiji, more than 600 farmers have participated in the production of Juncao mushrooms; while more than 1000 farmers participated in the cultivation of Giant Juncao grass for expanded livestock industry. The new Juncao ecological industry has become a characteristic industry opening up a new way for Fiji and other Pacific island countries to increase employment, reduce poverty, and cope with climate change and sustainable development.

中国援斐菌草技术示范中心
China-Fiji Juncao Technology Demonstration Center

第六章 服务国际减贫和可持续发展
Chapter 6　Promotion of International Poverty Reduction and Sustainable Develepment

2015年7月17日，斐济总理姆拜尼马拉马访问福建农林大学，种下象征中斐友谊的菌草
Fiji Prime Minister Hon. Josaia Voreqe Bainimarama visited FAFU, and planted Giant Juncao grass in the campus to symbolize the friendship between China and Fiji on July 17, 2015

2016年6月6日，斐济总统乔治·孔罗特夫妇视察中国援斐济菌草技术示范中心。他表示：菌草技术对斐济的贡献很大，不仅增加就业，提高农民收入，减少进口，也能解决畜牧业旱季饲料缺乏的问题
Fiji President H.E. George Khonote and First Lady visited the China-Fiji Juncao Technology Demonstration Center on June 6, 2016. He said: Juncao technology has made a great contribution to Fiji, which not only generates employment, increases income, and reduces imports, but also eases the shortage of feed in the animal husbandry during the dry season

2015年3月11日，福建省委常委、组织部部长姜信治（中）率福建省代表团访问斐济并到菌草技术示范中心看望菌草技术专家
On March 11, 2015, H.E. Jiang Xinzhi (middle), member of the Standing Committee of the Fujian Provincial Party Committee and Minister of the Organization Department, led a delegation from Fujian Province to visit China-Fiji Juncao Technology Demonstration Center

2015年7月4日，中国驻斐济大使张平陪同斐济总理姆拜尼马拉马视察中国援斐济菌草技术示范中心
Fiji Prime Minister Rt. Hon. Josaia Voreqe Bainimarama visited China-Fiji Juncao Technology Demonstration Center, accompanied by H.E. Zhang Ping, the Chinese Ambassador to Fiji on July 4, 2015

斐济时任农业部部长，现任外交部部长伊尼亚·塞鲁伊拉图在菌草技术示范中心亲手种植的巨菌草。自2014年以来，他共47次到中心调研指导
The Giant Juncao grass planted by H.E. Inia seruiratu Fiji's former Minister of Agriculture, now Minister of Foreign Affairs, China-Fiji in Juncao Technology Demonstration Center. Since 2014, he made 47 visits to the Center for inspection and guidance

2019年3月28日，中国农业农村部部长韩长赋在斐济农业部部长马亨德拉·雷迪博士陪同下视察菌草技术示范中心
On March 28, 2019, H.E. Han Changfu, the Minister of Agriculture and Rural Affairs of China, accompanied by H.E. Dr. Mahendra Reddy, Minister of Agriculture of Fiji, made an inspection visit to the Juncao grass plantation in China-Fiji Juncao Technology Demonstration Center

第六章　服务国际减贫和可持续发展
Chapter 6　Promotion of International Poverty Reduction and Sustainable Develepment

2019年3月30日，时任中国农业农村部副部长，现任联合国粮农组织总干事屈冬玉在钱波大使陪同下视察菌草技术示范中心

On March 30, 2019, Qu Dongyu, the former Vice Minister of Agriculture and Rural Affairs of China and the current Director-General of FAO, accompanied by H.E. Qian Bo the Chinese Ambassador to Fiji, inspected the China-Fiji Juncao Technology Demonstration Center

（1）中国援斐济菌草技术示范中心
China-Fiji Juncao Technology Demonstration Center

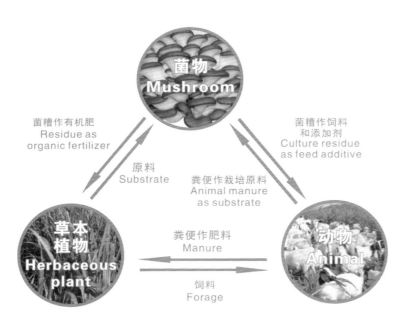

菌草新型生态产业循环图示
Diagram of Juncao's new eco-industry cycle

楠迪菌草标本园的巨菌草
Giant Juncao grass in Nadi Juncao Specimen Garden, Fiji

（2）菌草食药用菌生产
Production of Edible and Medicinal Mushrooms

工厂化出菇
Fruiting in factory

杧果树下栽培菌草灵芝
Cultivate Juncao *Ganoderma lucidum* under the mango trees

林下菌草栽培平菇
Pleurotus ostreatus cultivation under trees

(3) 菌草畜业
Juncao for Animal Husbandry

菌草饲料生产示范
Juncao feed production demonstration

菌草养羊
Raising goats with Juncao grass

（4）举办各种类型菌草技术培训班
Various Types of Juncao Technical Training Courses Held in Fiji

2014年11月6日，斐济首都苏瓦教会学校师生70多人到菌草技术示范中心参观学习
More than 70 teachers and students from Suva Missionary School made a study tour to the Juncao Technology Demonstration Center on November 6, 2014

2014年11月21日，福建农林大学党委书记叶辉玲给学员颁发结业证书
Ye Huiling, Secretary of the Party Committee of FAFU, delivered completion certificates to the trainees on November 21, 2014

第六章 服务国际减贫和可持续发展
Chapter 6 Promotion of International Poverty Reduction and Sustainable Development

2014年12月5日，斐济农业部部长率农业官员87人参观菌草技术示范中心
The Minister of Agriculture of Fiji led 87 agricultural officials made a visit to the Juncao Technology Demonstration Center on November 5, 2014

2015年3月11日，举办菌草技术骨干培训班
Training course for the technical backbones of Juncao industry was held on March 11, 2015

2015年4月9日，斐济南太平洋大学师生40人到菌草技术示范中心参观学习
40 teachers and students from University of South Pacific made a study visit to the Juncao Technology Demonstrtion Center on April 9, 2015

2015年4月21号，斐济辛阿托卡教会学校师生70人到菌草中心参观学习
70 teachers and students of Sigatoka Missionary School made a study visit to the Juncao Technology Demonstration Center on April 21, 2015

2015年6月10日，菌草菇烹饪实践
Practice on Juncao mushroom cuisines cooking on June 10,2015

2015年11月12日，斐济总理府办公室秘书长等18人到基地参观
The Principal Secretary with 18 officials from Fiji Prime Minister's Office made a visit to the Center on November 12, 2015

2018年8月31日，到斐济乡村举办小农户菌草技术培训班
The Juncao Technology Training Course was held for grass-root farmers in Fiji rural areas on August 31, 2018

6. 中非
Central African Republic

2018年，中非总统图瓦德拉出席中非合作论坛北京峰会，在与习近平主席会见期间，就发展包括菌草项目在内的农业合作达成共识。9月9日下午，图瓦德拉总统访问福建农林大学，考察了国家菌草工程技术研究中心，认为"发展菌草业可以帮助中非农民提高收入，解决饥饿问题，是中非脱贫和粮食安全的一条新途径"。

In 2018, H.E. Touadéra President of the Central African Republic attended the Beijing Summit of the China-African Cooperation Forum, and during his meeting with H.E. President Xi Jinping, they reached a consensus on the development of agricultural cooperation including the Juncao project. On September 9, H.E. President Touadéra made a visit to Fujian Agriculture and Forestry University for inspection of the National Juncao Engineering Technology Research Center. He commented that the development of Juncao industry could help the Central African farmers increase income and solve hunger problems, and it is a new way for Central Africa to alleviate poverty and enhance food security.

2018年9月9日，福建农林大学与中非共和国农业部签署开展菌草技术合作协议。

On September 9, 2018, Fujian Agriculture and Forestry University and the Ministry of Agriculture of the Central African Republic signed a cooperation accord to develop Juncao technology.

2019年3月，福建农林大学派出菌草技术专家访问中非，在中非总统府和国民议会举办菌草技术介绍会。中非总统图瓦德拉、国民议会议长恩贡、第一副议长马邦齐、国家经济计划部部长、农业部部长、畜牧业部部长、工商部部长、班吉市市长、班吉大学校长及83名议员等共400余人出席会议。

In March 2019, Fujian Agriculture and Forestry University sent Juncao technical experts to visit Central Africa and gave presentations on Juncao technology in the Central African Presidential Palace and the National Assembly. There were more than 400 people including Central African President H.E. Touadéra, National Assembly Speaker H.E. Ngon, First Deputy Speaker, the Minister of National Economic Planning, the Minister of Agriculture, the Minister of Animal Husbandry, the Minister of Industry and Commerce, the Mayor of Bangui, the President of Bangui University and 83 members of Parliament attended the meetings.

2019年6月17日至8月5日，中非首批25位学员到福建农林大学学习菌草技术。7月12日，福建省委书记于伟国、副省长郭宁宁到福建农林大学看望研修班学员，并与学员座谈。于伟国书记表示对中非引进菌草技术的支持："福建省将继续加大支持力度，更好地推动菌草技术在包括中非在内的发展中国家或欠发达国家落地生根，为构建更加紧密的中非命运共同体作出更大贡献。"他还说："无论你来自哪里，只要提到菌草，我们就是好朋友！"

From June 17 to August 5, the first batch of 25 participants from the Central African Republic went

Chapter 6 Promotion of International Poverty Reduction and Sustainable Develement

to Fujian Agriculture and Forestry University to attend Juncao technology training seminar. On July 12, Yu Weiguo, Secretary of the Fujian Provincial Party Committee, Vice Governor of Fujian Province, Guo Ningning made a call on the trainees of Central Africa in Fujian Agriculture and Forestry University, and had a discussion with the trainees. H. E. Yu Weiguo emphasized his support: Fujian Province will continue to increase support to better promote Juncao technology to take root in developing or underdeveloped countries, including Central Africa, and make greater contributions to building a closer China-Africa community with a shared future. He added that "No matter where you are from, as long as you are working on Juncao, we are good friends!"

2020年1月24日，中国政府把菌草技术列为援中非技术项目。现已建立菌草技术示范基地和菌草技术扶贫示范村，为中非增加就业、解决饥饿、减贫、维护社会稳定、促进经济发展走出一条新路。

On January 24, 2020, the Chinese government listed Juncao technology as a China aid project to Central Africa which up to now has established Juncao technology demonstration base and Juncao technology poverty alleviation demonstration village, facilitated a new path for Central Africa to generate employment, eliminate hunger, reduce poverty, maintain social stability, and promote economic development.

2018年9月9日，中非总统福斯坦·图瓦德拉访问福建农林大学
On September 9, 2018, H.E. President Faustin Touadéra of Centrl African Republic made a visit to FAFU

中非总统福斯坦·图瓦德拉在福建农林大学国际菌草苑种植菌草纪念

H. E. President Faustin Touadéra of Central Africa plants Giant Juncao grass in the International Juncao Garden in FAFU

在总统府的介绍会上，图瓦德拉总统表示菌草技术为中非农业发展开辟了一条全新的道路

At the presentation meeting in the Presidential Palace, H.E. President Touadéra said that Juncao technology has opened up a brand-new path for the development of agriculture in Central Africa

恩贡议长表示中非国民议会议员将全力推动菌草技术在各自选区普及落地，造福中非人民，希望两国开展更多有利于提高百姓福祉的务实合作

H.E. Speaker Ngong said that members of the National Assembly of Central Africa should make every effort to promote the popularization of Juncao technology in their respective constituencies and benefit the people of Central Africa. He hoped that the two countries would carry out more practical cooperation to improve the well-being of the people

第六章 服务国际减贫和可持续发展
Chapter 6　Promotion of International Poverty Reduction and Sustainable Develepment

中非总统图瓦德拉和中国驻中非大使陈栋一起在中非种下了首批巨菌草和菌草平菇
H.E. President Touadéra and China's Ambassador to Central Africa H.E. Chen Dong planted the first batch of Giant Juncao grass and oyster mushroom in Central Africa

2019年6月17日~8月5日，中非首批25位学员在国家菌草中心学习菌草技术。福建省委书记于伟国（前排右七）、副省长郭宁宁（前排右四）看望学员，并与他们进行了座谈
From June 17 to August 5, 2019, the first batch of 25 participants from Central Africa came to National Juncao Research Center to learn Juncao technology. Yu Weiguo (front row seventh from right), Secretary of the Fujian Provincial Party Committee, and Guo Ningning (front row fourth from right), Vice Governor of Fujian Province, called on the trainees and had a discussion with them

2019年7月11日，中国国家国际发展合作署署长王晓涛（二排左五）看望中非培训班学员
On July 11, 2019, H. E. Wang Xiaotao, Director General of China National Agency for International Development Cooperation (second row, fifth from left), called on the participants of the training course for Central Africa

福建农林大学校长兰思仁为学员颁发结业证书，并转交于伟国书记为学员准备的纪念品
Prof. Lan Siren, President of FAFU delivers Completion Certificates to the trainees and handed over to them the souvenirs prepared by H.E. Yu Weiguo

2019年8月22日～9月20日，在班吉举办中国援中非菌草技术海外培训班。中国驻中非大使陈栋、中非农业及农村发展部部长费祖雷出席开班仪式
From August 22 to September 20, 2019, the China Aid Juncao Technology Overseas Training Course in Central Africa was held in Bangui. H.E. Ambassador Chen Dong and Minister of Agriculture and Rural Development H.E. Feizoure of Central Africa attended the opening ceremony

第六章 服务国际减贫和可持续发展
Chapter 6 Promotion of International Poverty Reduction and Sustainable Development

2019年12月1日，中非总统图瓦德拉在61周年国庆日上为林占熺颁发"中非国家感恩指挥军官勋章"，为林冬梅、林辉二人颁发"中非国家感恩军官勋章"，为蔡杨星、罗德金、祝粟三人颁发"中非国家感恩骑士勋章"。中非共和国总理、议长，非洲多国政要，各国际组织代表，各国驻中非外交使节等千余人出席了此国庆授勋仪式

On December 1, 2019, H. E. Touadéra, President of the Central African Republic, awarded Prof. Lin Zhanxi the "National Medal of Gratitude to Commanding Officer", Dr. Lin Dongmei and Mr. Lin Hui the "National Medal of Gratitude to Officer", Mr. Cai Yangxing, Mr. Luo Dejin and Mr. Zhu Su the "National Medal of Gratitude to Knight" on the 61st National Day. More than 1000 participants attended the awarding ceremony, including the Prime Minister, and the Parliament Speaker, the government officials from other African countries, representatives of international organizations, and diplomatic envoys from various countries in Central Africa Republic

中国—联合国和平与发展基金菌草技术重点项目

Juncao Technology Key Project of the China-United Nations Peace and Development Fund

20 多年来,菌草技术在发展中国家应用的实践证明,对实现联合国 2030 年可持续发展议程 17 项目标中的 13 项都可起到促进作用。

For more than 20 years, the application of Juncao technology in developing countries has shown that it can promote 13 out of the 17 goals of the United Nations 2030 Agenda for Sustainable Development.

菌草技术援外项目实施的重要目标是增加就业和减贫,逐步拓展到 13 项可持续发展目标。通过菌草技术与菌草产业扶贫,从资源、技术、人才到产业链,实现原料资源可持续利用、土地资源可持续利用、保护生态可持续、经济发展可持续、本土人才培养可持续,形成了一个全面的解决方案,从而破解受援国的发展难题。

The key target in the implementation of the Juncao technical foreign aid project is to increase employment and reduce poverty, and now gradually expanded to 13 sustainable development goals. Through development of Juncao technology and Juncao industry, successful poverty alleviation models integrate resources, technology, talents to the industrial chain, so as to achieve sustainable use of raw material resources, and to achieve all round sustainable development in utilization of land resources, ecological protection, economic development, and local talent training. Thus to develop a comprehensive solution to overcome the development constraints of the recipient countries.

联合国经社部、联合国粮农组织、世界粮食计划署等联合国机构积极支持菌草技术推广,促进菌草技术在落实联合国 2030 年可持续发展议程作出新的实质性贡献。

The UNDESA, UNDP, FAO together with other UN agencies have always actively supported the promotion of Juncao technology, for substantive contributions to the implementation of the United Nations 2030 Agenda for Sustainable Development.

2017 年,联合国经社部把菌草技术列入中国—联合国和平与发展基金重点推进项目。

UNDESA in 2017 listed Juncao technology as a key promotion project to be implemented under the China-United Nations Peace and Development Fund.

Chapter 6 Promotion of International Poverty Reduction and Sustainable Development

菌草技术服务联合国2030年可持续发展议程13项目标

Juncao technology promotes the implementation of 13 goals of the United Nations 2030 Agenda for Sustainable Development

2017年5月26日，在纽约联合国总部召开了"中国—联合国和平与发展基金菌草技术项目研讨会"。

China-United Nations Peace and Development Fund Juncao Technology Project Seminar was held at the United Nations Headquarters in New York on May 26, 2017.

这一项目紧扣发展中国家普遍关心的消除贫困、减少饥饿、可再生能源利用、促进就业和应对气候变化等问题，结合非洲、亚洲国家具体国情和需要，积极贡献"中国方案"，帮助非洲、亚洲等发展中国家破解发展难题，落实可持续发展目标，推进全球发展事业。

——2017年5月26日，常驻联合国代表刘结一大使在"中国—联合国和平与发展基金菌草技术项目研讨会"上发言

This project closely follows the issues of poverty eradication, reduction of hunger, use of renewable energy, employment promotion, and response to climate change that developing countries are generally concerned about. It combines the specific national conditions and needs of African and Asian countries and actively contributes to the Chinese Program to help Africa , Asia and other developing countries solve development problems, implement sustainable development goals, and promote global development.

—— Ambassador Liu Jieyi, Permanent Representative to the United Nations Speech at the Peace and Development Fund Juncao Technology Project Seminar on May 26, 2017

第六章 服务国际减贫和可持续发展
Chapter 6　Promotion of International Poverty Reduction and Sustainable Develepment

菌草技术发明人林占熺作主旨发言介绍菌草技术及其在发展中国家的应用
Lin Zhanxi, the inventor of Juncao technology, made a keynote presentation to introduce Juncao technology and its application in developing countries

2017年9月9日，在中国鄂尔多斯召开《联合国防治荒漠化公约》第13次缔约方大会菌草技术边会
Juncao Technology Side Event on 13th Session of the Conference of the Parties to the United Nations Convention to Combat Desertification was held on September 9, 2017 in Ordos, China

2018年6月4~5日,在斐济楠迪召开菌草技术能力建设区域研讨会——助推太平洋小岛屿发展中国家发展可持续农业,实现可持续发展目标

Regional Capacity Building Workshop on Juncao Technology and its Support to Achieve Sustainable Agriculture and the Sustainable Development Goals for Pacific Small Island Developing States was conducted on June 4-5, 2018 in Nadi, Fiji

2019年2月14~18日,在老挝万象召开老挝菌草技术能力建设咨询会——助推农业可持续发展、实现可持续发展目标

Juncao Technology Capacity Building Advisory Trip to the Lao People's Democratic Republic for Achieving Sustainable Agriculture and the Sustainable Development Goals was convened on February 14-18, 2019 in Vientiane, Laos

第六章　服务国际减贫和可持续发展
Chapter 6　Promotion of International Poverty Reduction and Sustainable Develepment

2019年4月18日，在纽约联合国总部召开联合国菌草技术高级别会议。

On April 18, 2019, the United Nations Juncao Technology High-level Conference was held at the United Nations Headquarters in New York.

通过菌草技术，中国给我们讲了一个伟大的故事，这个故事现在已经分享到100多个受益于这一创新的国家。在福建省点燃的火花已经显示了一个创新的潜力，如果将其善加培育和部署得当的话就能改变世界各地人们的生活状况和改善他们的生计。

——2019年4月18日，
第73届联合国大会主席玛丽亚·费尔南达·埃斯皮诺萨·加西斯
在联合国菌草技术高级别会议上致辞

Through Juncao technology, China has a great story to tell—a story now stared with over 100 countries who have benefited from this innovation. The spark lit in Fujian Province has shown the potential of a single innovation—if nurtured and deployed wisely—to change lives and improve livelihoods across the world.

——Statement by H.E. Mrs. María Fernanda Espinosa Garcés, President of the 73rd Session of the UN General Assembly, on April 18, 2019

菌草技术是一项着重扶贫、保护生态和促进可持续发展的综合性技术，已在众多"一带一路"沿线国家实施，助力有关国家落实消除贫困和饥饿、促进粮食安全、保障和增加就业、应对气候变化和保护生态环境等多项可持续发展目标，帮助有关国家摆脱了贫困，过上了更加美好的生活。

——2019年4月18日，
中华人民共和国常驻联合国代表马朝旭大使
在联合国菌草技术高级别会议上讲话

Juncao technology is a comprehensive technology that focuses on poverty alleviation, protection of the ecology and promotion of sustainable development. It has been implemented in many countries along the "Belt and Road" to help relevant countries implement poverty and hunger elimination, promote food security, ensure and increase employment, a number of sustainable development goals such as tackling climate change and protecting the ecological environment have helped relevant countries get rid of poverty and lead a better life.

——Ambassador Ma Zhaoxu, Permanent Representative of the People's Republic of China to the United Nations,
Speaking at the UN Juncao Technology high-level Conference on April 18, 2019

2019年4月18日，福建省副省长郑建闽代表福建省人民政府在菌草技术高级别会议上致辞
On April 18, 2019, Zheng Jianmin, Vice Governor of Fujian Province, delivered a speech at the UN Juncao Technology High-level Conference on behalf of the Fujian Provincial People's Government

2019年4月18日，福建省副省长郑建闽（中）、福建农林大学校长兰思仁（右）参观在纽约联合国总部展出的菌草技术图片展
On April 18, 2019, Zheng Jianmin, Vice Governor of Fujian Province (middle), and Prof. Lan Siren, President of FAFU (right), visited the Juncao Technology Photo Exhibition at the United Nations Headquarters in New York

2019年4月18日，福建农林大学校长兰思仁（右三）在联合国菌草技术高级别技术磋商会议上致辞
On April 18, 2019, Prof. Lan Siren (third from right), President of FAFU, delivered a speech at the UN Juncao Technology High-level Conference

第六章　服务国际减贫和可持续发展
Chapter 6　Promotion of International Poverty Reduction and Sustainable Develepment

2019年4月19日，菌草技术发明人林占熺在菌草技术高级别会议上作《发展菌草业，造福全人类》的主旨发言

On April 19, 2019, Juncao technology inventor Lin Zhanxi made a keynote speech at the UN High-level Conference on *Develop Juncao Industry for the Benefit of Mankind*

2019年6月18~21日，菌草技术支持联合国2030年可持续发展目标高级别考察团

International Study Tour on Juncao Technology and its Support to Achieve Sustainable Agriculture and the SDGs on June 18-21, 2019, in Fuzhou, China

2020年1月7~11日，在马达加斯加召开政策制定者和农户能力建设研讨会——菌草技术加快推进2030年可持续发展议程

Capacity Building Workshops for Policymakers and Farmers on Juncao Technology to Accelerate Progress in the 2030 Agenda for Sustainable Development was held in the Republic of Madagascar on January 7-11, 2020

2021年7月19~22日，世界粮食计划署（WFP）驻华办公室代表屈四喜（右）调研内蒙古、宁夏菌草科技创新产业园

Qu Sixi (right), Resident Representative of WFP's China Office, made a study visit to Juncao Science and Technology Innovation Industrial Parks in Inner Mongolia Autonomous Region and Ningxia Hui Autonomous Region on July 19-22, 2021

尼泊尔国家科学技术院与国家菌草中心合作，在低海拔与高海拔省份设立菌草研究与推广基地

Nepal Academy of Science and Technology cooperates with National Juncao Research Center to establish Juncao Research Extension Bases at low altitude and high altitude areas

Chapter 6 Promotion of International Poverty Reduction and Sustainable Development

老挝农业部为解决饲草短缺引进菌草，已设立菌草种苗基地，为全国各地提供草种
Laos Ministry of Agriculture has set up Juncao breeding base to supply Juncao grass seedings all over the country

国家菌草中心与埃及沙漠研究所、艾资哈尔大学及农业部食品科学研究所合作，联合培养菌草专业的研究生
National Juncao Research Center cooperated with Egypt Desert Research Institute, Al-Azhar University and Food Science Research Institute of MOA to jointly cultivate postgraduates in the research field of Juncao

朝鲜农业部科技人员在平壤菌草基地学习巨菌草种苗越冬技术
The reseachers from the Ministry of Agriculture of DPRK learned the overwintering technology of Giant Juncao grass seedlings at the Pyongyang Juncao base

中国专家在厄立特里亚示范基地指导菌草种植与栽培食用菌
Chinese expert gave guidance on Juncao planting and mushroom farming at the demonstration base of Eritrea

第一期萨摩亚菌草技术培训班在线举办，学员与国家菌草中心教师远程合影
The 1st Juncao Technology Training Course for Samoa was held online. Participants had group photo with lecturers of National Juncao Research Center in China

第六章 服务国际减贫和可持续发展
Chapter 6　Promotion of International Poverty Reduction and Sustainable Development

包奇州政府支持设立中国-尼日利亚菌草技术示范点
Bauchi state supports the construction of China-Nigeria Juncao Technology Demonstration Site

南苏丹农业咨询组织建立菌草种苗圃，推广牧区农户种植菌草饲喂家禽家畜和养鱼
South Sudan Agriculture and Advisory Organization has set up Juncao Nursery and extended Juncao grass planting in pastoral communities for feeding poultry, livestock and fish

李玉春（沙巴州农村发展机构项目经理），她在沙巴进行试验和推广菌草技术，并建立菌草菇的示范基地，成为当地农村扶贫发展的重要项目，促成国家菌草中心与沙巴州农业厅的合作
Lee Nyuk Choon(Project Manager of Sabah Rural Development Agency).She conducted trials and promotion in Sabah and set up the Juncao demonstration base. The project has become the primary project for poverty alleviation and development in local rural areas. She facilitated the cooperation between the Ministry of Agriculture of the Sabah State and FAFU National Juncao Research Center